Compound Polyhedra

Compound Polyhedra

Modular Origami

Fergus Currie

 Tarquin

About the author

Fergus Currie was born in Edinburgh, Scotland, but now lives and works in Athens, Greece. He studied composition at the Royal Academy of Music in London and the University of Birmingham where he received an MA for his research into the work of Japanese composer Toru Takemitsu. He has worked as a professional musician for more than 35 years and has been a member of the orchestra of the Greek National Opera since 2009.

© Fergus Currie 2021
ISBN Paperback: 978-1-913565-72-5
ISBN Cased: 978-1-913565-73-2
ISBN EBook: 978-1-913565-74-9
Designed by Versatile PreMedia
Printed in Europe
Distributed in the USA by IPG: www.ipgbook.com

Tarquin
Suite 74, 17 Holywell Hill
St Albans AL1 1DT
United Kingdom
www.tarquingroup.com

Contents

Introduction

During the COVID-19 outbreak in 2020, I found myself, as did millions of other people around the world, confined to my home for an extended period of time. The idea of weeks or even months at home seemed terrifying but it quickly dawned on me that this was a great opportunity to organise my origami designs (which, until then, had been either jotted on scraps of paper or, even worse, confined to memory) into a study of carefully drawn diagrams and instructions.

As a teacher of music composition, I tell my students to present their work in such a way that if they were to put their score in a box and sent it to another planet, someone there could perform it perfectly, just from the instructions on the score. Although I do follow my own advice with music, my approach to origami design was clearly very different and that had to change. So, in April 2020 with the COVID-19 lockdown underway, I tentatively began preparing proper folding diagrams for some of my models. I didn't have a professional CAD programme on my computer, so I worked with the graphics section of my word processor, dragging lines and boxes around, changing line thicknesses and filling various areas with colour until I eventually got the hang of this rather unwieldy tool. The result is this collection of ten complete designs.

The book concentrates on one specific type of geometric form, the regular compound polyhedra (forms where two or more regular polyhedra are integrated into one model). Only the last two models are not of this type – the Small Triambic Icosahedron (STI) is given so that the reader may better visualise the Compound of the Dodecahedron and the STI, the irregular compound polyhedron found in the final chapter.

The majority of these models are for the experienced origami folder and should probably not be attempted by absolute beginners. Many of them use advanced techniques either in folding or assemblage and I strongly recommended that readers make several extra units for 'trial runs' before embarking on what is to become their finished model.

The models are arranged roughly in order of complexity, which happily coincides with the level of folding difficulty, so beginners and intermediate folders should start at the beginning but more advanced folders may feel free to jump in anywhere.

The construction of several of the models involves pre-folding templates, the technique for which i s described in the instructions. Care should be taken to fold these tools correctly since flaws in a template are carried into every unit of a model and are thus amplified by repetition unlike flaws in a single unit.

Many of the designs rely on the technique of folding between two reference points, which means that folds are made to connect intersections of existing folds. Some of the trickier ones are marked in the diagrams with small circles on the points involved.

All of the prototypes for this book were made from sheets cut from commercially available coloured A4 80 g/m^2 printer copy paper. It should be noted that some papers with the same nominal weight may have a better 'memory' and be more resilient than others. It makes sense to organise your paper stock in terms of these attributes and to find colour matches within similar types, otherwise your model – especially the more complex ones – might end up looking a little the worse for it.

I dare say there is enough material in this book to keep a dedicated folder busy for a good few weeks but I suggest that you leave and come back to these models from time to time, enjoying their individual forms and idiosyncrasies for a while before moving on to another model. In this way modular origami becomes something like a form of meditation. The more one studies the forms of geometric models, the deeper one's understanding of the interconnectivity of all things.

The Art of Geometry in Origami

Compounds of regular polyhedra, as mentioned above, are forms that combine two or more regular polyhedra. There, too, are other restrictions that define such compounds. The first is that there should be a regular and harmonic intersection of the edges and/or vertices. The second is that the positioning of each polyhedra must be described as a repeated regular rotation about a common axis. In many cases, these axes are easy to describe in mathematical terms but in others, they may require complex expressions of irrational numbers such as $\sqrt{2}$, $\sqrt{5}$, π, ϕ, etc. The first chapter, for example, describes how to build the Compound of Two Cubes (arguably the simplest polyhedron compound), where the position of the second cube can be described as a rotation of the first cube by 60 degrees about an axis of its internal diagonal. After making the model, the reader might examine this by holding up the carrousel vertices as North and South poles and rotating it slowly. The Compound of Three tetrahedra has a rotation of 60 degrees between its elements, while the Compound of Three Octahedra has a rotation of 120 degrees between its elements.

One of the most problematic tasks in designing mathematical modular origami is finding elegant approximations for the exact (desired) angles in geometric forms that would otherwise require excessively convoluted folding to reproduce. The two main considerations are to find a balance between simple origami and minimal divergences from the desired angle. Much effort has been made here to ensure geometric accuracy, but a few of the more complex models use very close approximations, most of which lie within 99% of the desired angles. Origami, then, is surely an art form.

That interesting mixture of mathematics and aesthetics rings comfortably in my musician's ears, recalling perhaps J. S. Bach's fugues in the simplicity of their starting material, and how, through repetition and variation, they create an endless cornucopia of enchanting and beautiful geometrical wonders.

Fergus Currie

Athens, February 2021

Key to the Diagrams

The diagrams that comprise the main part of this book employ a system of symbols and technical terms, which are explained below. For each model, they are presented in sequence with concise instructions directly below.

Paper

Initially, the paper is presented reverse side up, with the coloured outer side face-down. The edge of the paper is always represented by a bold line even when it is folded over. Feel free to move the paper to make it easier to fold.

reverse side of the paper

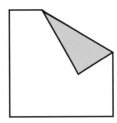

a fold reveals the outer side of the paper

Folds, Creases and Arrows

A **fold** involves one movement whereas a **crease** is the result of folding and unfolding. A fold or a crease is shown completed with a bold dotted line. The initial diagram also shows a faint impression of the paper's position before the fold. Subsequently the completed fold is shown as an unbroken bold line, while the creases is shown as a thin line.

a fold being made

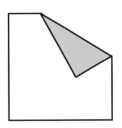

the same fold completed

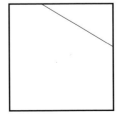

the crease after unfolding

9

An **arrow** is used to show whether the manoeuvre creates a fold or a crease. A fold is accompanied by a **single arrow**, while a crease is accompanied by a **double arrow**.

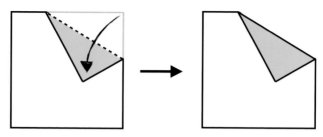

single arrow for a fold

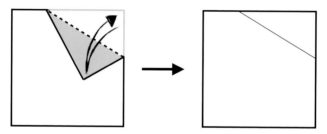

double arrow for a crease

Very occasionally the paper will not be folded to the front, but to the back in what is traditionally known as a **mountain fold**. This is shown as a line of alternate dots and dashes.

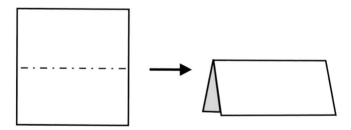

the result of a mountain fold

Reference Points

A **reference point** is any point on the paper that is used as a guide for making a subsequent fold or crease. These might include a corner, a point where a crease meets the edge of the paper, the intersection of two creases, or marks made by folding a fraction of the length of an edge. As you can see, there are many different cases but a couple of examples should suffice to clarify the meaning.

If we fold the paper in half horizontally, then use the place where that crease meets the left edge as one reference point and the top-right corner as another reference point, the crease connecting these two points will be at an angle of 26.6 degrees with length $\frac{\sqrt{3}}{2}$ times the edge of the square.

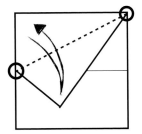

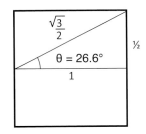

making a crease between two reference points · the resulting angle $\tan^{-1} 0.5 = 26.6$

Reference points may also be abstract. Let's fold a square in half horizontally, then bring the bottom-right corner about the fulcrum of the top-right corner until it falls on the horizontal crease. You can see that the point where the corner falls on the centre line was not discernible before. It is deduced or 'abstracted' from the process.

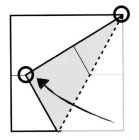

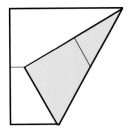

using an abstract reference point

Tools, Tips and Techniques

Before beginning any of the models, make sure you have the following tools at hand. Some are for cutting the paper to the initial sizes; others are for helping to fold the modules and others are for manipulate the modules during the assembly stage. Your tools may not be exactly the same as mine, but if they can perform the tasks asked of them then they will do just fine.

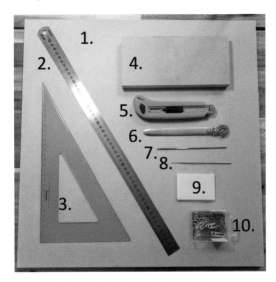

Fig. 1 all the tools used in this book

1. 50 × 50 cm folding board (a base for folding the modules)
2. steel ruler (used for cutting the initial paper sizes)
3. set-square (used for cutting the initial paper sizes)
4. high folding-block (used occasionally to fold through multiple layers)
5. craft knife (used for cutting the initial paper sizes)
6. letter opener (used to make crisp folds and creases)
7. 2.5 mm crochet hook (used for manipulating the inside of the model)
8. 1.2 mm crochet hook (used for manipulating the inside of the model)
9. sticky memos (used to support the model during assembly)
10. paper clips (used to support the model during assembly)

At least one strong moveable angle-poise lamp shining slightly diagonally across the folding board will help define any creases that may be used as references. You will also find various 'Top Tips' dotted throughout the book to help you get the best possible results. However, some techniques need to be explained at the start. These may already be known to the experienced folder, but for anyone new to modular origami, the following sections should be studied carefully before attempting to make the models in this book.

Creasing and Folding Tips

- When you make any crease or fold, you should be totally aware of where it starts and finishes.
- Try not to crumple the paper when first bringing the parts of the fold together.
- Press the fold or crease gently into place before smoothing with the bone.
- Use the flat of the bone to press out the fold, not the sharp edge.
- If the fold has a narrow end, align that first then the broad end.
- Only fold/crease when you are certain you have it aligned correctly.
- If there are multiple layers to fold through, you may use the bone to keep the inside folds in place as you prepare it.
- Work slowly and accurately.

Assembly Tips

- Never begin the final assembly of any model before you have all of the modules ready.
- Follow the instructions. The order in which the modules are put together may often be crucial to the model's success.
- Have paper clips and sticky memos at the ready. I cut off most of the non-sticky part of the memos when I open the pack, leaving just enough to get a hold of them when they need to be removed.
- Make a few extra modules, especially for the more difficult models, to learn how the joints work. Very often in modular origami, you are required to make some of the last joints 'blind' on the inside of the model, so a good understanding of how the joints work, along with some practice, goes a long way to making a good model.
- Have patience! Treat the assembly stage like one of those mind-boggling Chinese woodblock puzzle. Don't succumb to the urge to plunge your fist through an almost completed model in frustration. Rather go and have a cup of tea or a walk, then come back to it with a clear head and figure out what needs to be done, and do it.

Individual Techniques

In order to explain the various individual techniques, this section follows the making of a simple octahedron model step by step, with each step explained in full. In the main chapters, the instructions given are more concise and assume that the folder already understands the following techniques.

1. **Cutting the Paper to Size**

   You will need four square sheets of 80 g/m^2 paper for this model. Here's an easy way to cut accurate squares from A4 or US Letter or any other rectangular paper size.

1. Line up several sheets and place them on your cutting board. Place another sheet of the same size perpendicularly on top of them, lining up the right edge with a set-square. Holding everything in place, set your steel ruler along the left edge of the top sheet and on top of the sheets you are going to cut.

2. Remove the top sheet and cut along the edge of the ruler.

3. This should give you a set of identical squares for your model. Do not try to cut more than six squares at once since the ruler might slip before you cut through all the sheets.

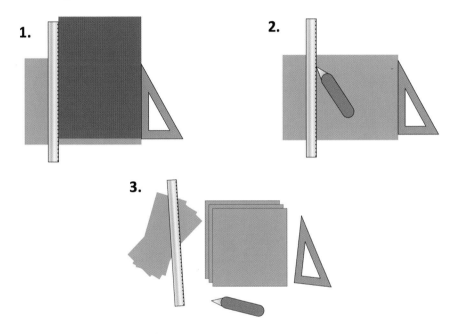

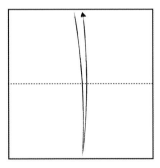

2. Corner Alignment

The first folding technique helps make an accurate fold along the vertical centre line of the module. Take the bottom edge of the paper and gently roll it up so that it almost matches the top edge. Do not fold at this stage. Softly slide the upper layer at one corner so that the lower layer just disappears underneath it. Place a fingertip on this to hold it in place, but not so hard that the paper cannot slip slightly, then align the other corner in the same way. You may need to make a slight adjustment to the first corner. When you are satisfied that both corners are aligned, tightly hold the top edges in place with two fingers and smooth the middle part downwards from the top with the other hand. When you have made the initial fold along much of the centre line, press out the ends while securing the corner alignments. Lastly, as with all folds you will make, gently 'iron' the entire fold with the flat of a letter opener (or an origami bone).

3. Reference Creases

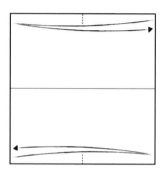

Making reference creases involves aligning certain features and creasing a small portion so that the specific point may be used in a future fold or crease. In this case, the certain features are the top-left and top-right corners. By aligning them as above, the top edge is now pressed flat at the midpoint. The same process is then applied to the bottom edge. These short creases become the reference points for the following fold.

4. Folding to a Reference Point

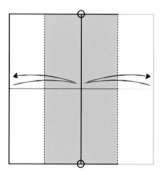

When reference points are used to make a fold, the alignment should exactly upon the reference. The reference creases made above should almost disappear under the folded portion. Again, make sure of all the alignments before making the crease. In our example we will fold the left and right edges inwards so that meet the reference creases. The subsequent crease will divide the module vertically into three parts – one quarter, one half, and one quarter.

5. Folding to an Abstract Reference Point

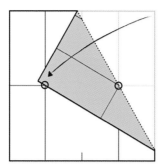

In our example, we will fold diagonally from the intersection of the horizontal centre line and the right horizontal crease. To obtain the correct angle for the fold we must first fix the intersection and align one abstract reference point, using the other intersection. Bring the top right corner towards the middle and observe how the intersection reacts. At some point, the paper will buckle upwards, forming an apex at the intersection. Rotate the right edge so that it falls on the other intersection, while maintaining a clean apex at the right intersection. When you have both of these points aligned, you can press down the paper to make a diagonal fold, which falls on the right intersection.

6. Mirroring

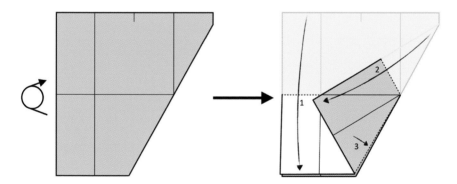

Mirroring a crease involves making a copy of it on another part of the module by using another crease as a virtual mirror. In this case the mirror is going to be the horizontal centre crease and we will make a reflection of the diagonal crease we've just made. To begin with, flip the module over as indicated by the looped arrow. You will see that, on either side of the horizontal centre line, there is a larger and a smaller section. We need to lift up the top layer of the larger section up and begin to fold along the longer section of the horizontal centre line (1). The bottom layer of this section is folded along the shorter portion of the horizontal centre line, bringing the top layer over (2). Align the diagonal edges and press the fold (3). When you unfold everything, you will see that the new fold is a mirror image of part of the diagonal fold in the 'mirror' of the horizontal centre line.

7. Extension of a Partial Crease

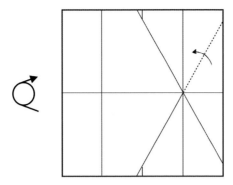

Unfold everything and return the module to its original position. The next step is to extend the new crease to the edge of the module. In the diagrams, extensions are not shown as folds or creases, only the extension of the existing crease is shown with an arrow straddling it. refold the existing crease and continue it as far as is shown in the diagram. More often than not this will be to the edge of the module.

Once you have completed the extension, rotate the module 180 degrees and repeat steps 5, 6 and 7 for the other half. These steps will usually be fully diagrammed but occasionally, in simpler designs such as this one, a direction to repeat certain steps for the other side is given.

8. The Inside Reverse Fold

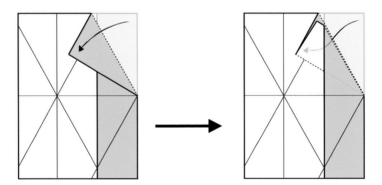

An inside reverse fold is made when a fold is made through two layers and the then rearranged so that the folded part rests in between the top and bottom layers. This is done by 'reversing' the direction (i.e. mountain folds become valley folds and vice versa) of the top two folds. Often this kind of fold is made without making the first fold shown above as in the second diagram.

9. Partial Folds

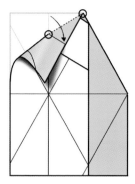

A partial fold (or crease) is one that is not creased all the way to the edge of the paper. There are many such folds in this book and the points where the partial folds begin and end are marked with small circles. Also, many of the diagrams involving partials have shading or a three-dimensional appearance to them to help clarify their limits. In this case the fold is made only between the top edge and the left vertical crease, having first aligned the top edge of the module with the top of the inside reverse fold made in step 8.

10. The Swivel Fold

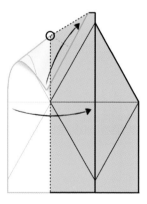

A swivel fold involves rotating two parts of the module about a common fulcrum. The fulcrum is shown by a small circle and the arrows show the two complimentary folds. When making swivel folds you should be aware of the fulcrum at all times so that it doesn't slip before pressing down any final folds. Align both parts accurately and gently press the final fold starting at the fulcrum. In this example, bring the top left corner of the module, which was raised in step 8, up to the top edge, along the existing crease. At the same time bring the left quarter of the module over to meet the vertical centre line. Press out the fold at the top-left of the module.

11. Tucking Inside

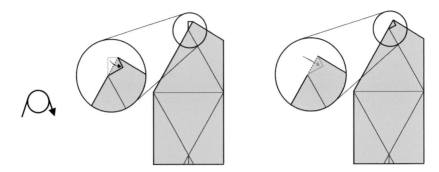

Many of the diagrams uses a 'magnifying glass' type zoom to show smaller details of the folding sequence. In the above diagram, the module has been flipped horizontally and we have zoomed in on the upper tip to show the detail. Here, the tip is folded over onto the module, then it is tucked inside the top layer. Once you have completed the tuck, you should unfold the module, rotate it by 180 degrees and repeat steps 8-11 for the other half.

12. Collapsing the Module

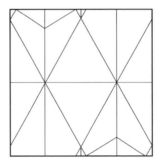

By now, your module should resemble the 'crease pattern' in the above diagram. The directions of the existing creases should facilitate the next step, which is to 'collapse' the module into its flat form.

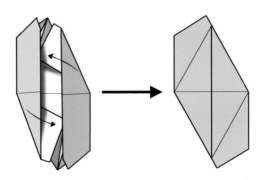

Refold steps 8-11 but let the folds open up slightly. Now refold the repeat of steps 8-11 you did for the other side, placing the inside reverse fold under the bottom end of the swivel fold made in step 10. Do exactly the same thing for the other half of the module. All of these things need to happen together.

13. Folding Through Multiple Layers

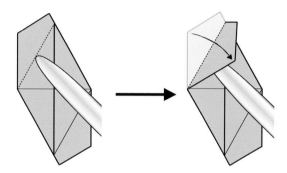

When folding through several layers of paper, you should use a blunt edge (anything from a spoon handle to a commercial origami bone will do) to hold the inner layers in place as you go. Ensure that the fold is flat on the folding board, then put the blunt edge on the fold to hold it down then gently bring the free side upwards. Move the blunt edge along the fold to accommodate the folding. Complete the fold with the blunt edge still in place, then pull it out and use it to press the fold flat. When you have completed the fold in the diagrams above, rotate the module 180 degrees and make the same fold at the other end. Your result should look like the diagram below.

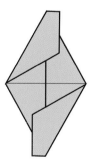

14. Folding Round the Module

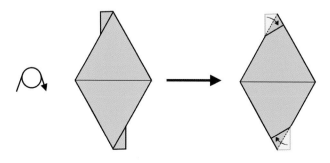

The final folds made on this module are made the same way that the first part of the tuck in step 11. The module is flipped over horizontally and the extruding portion of the fold made in step 13 is folded round on to the front. Try to keep the actual fold as close to the edge of the module as possible. Gently ease open the folds made here and in step 13. The finished module should resemble the diagram below.

You will need to make another three identical modules before moving on to the assembly process. Keep your modules in a safe place and don't be tempted to 'try out' the joints before you have made all four of them.

15. Assembly Procedure

For each model the assembly instructions are different. You should read through them a couple of times and make sure you understand what needs to be done, and in what order. It would be a shame to have made 48 modules only to crumple them in a rushed attempt to assemble the model. That said, you can always fold a new module if one gets torn or crumpled too much. The example here is not too tricky but, as with most modular origami models, the final closing of the model requires some gentle persuasion. The diagrams are meant to guide you through the process but you should also treat the assembly a bit like a Chinese woodblock puzzle, where pieces fit together only when combined in the exactly right order and angle. The watchword here is patience.

The first of the assembly diagrams usually explains the joints. Here, we have a simple tab and slot arrangement.

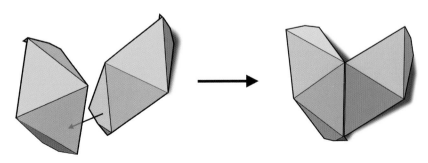

Next, connect the four modules together at one apex.

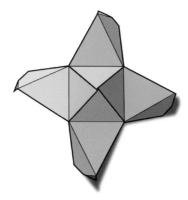

Finally, match all four tabs and slots of the opposite apex and gradually close the model. This is where it begins to feel like one of those Chinese puzzles. Makes sure that both parts of all the tabs are inside their respective slots – this might be done with the aid of a medium (2.5 mm) crochet hook or small knitting needle if need be.

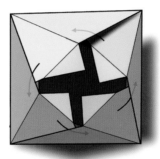

When the joints begin to slide into place, gently massage the model into shape, checking that none of the tabs have been deformed as they were pushed into place. Below is a photograph of the finished Octahedron model.

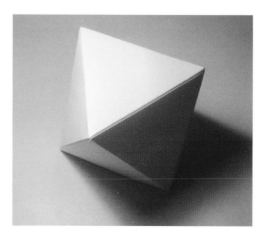

Now that we have seen most of the basic techniques in detail, you should be ready to tackle the compound polyhedra models in this book. They are arranged roughly in order of difficulty and each has a rating underneath the title. So, arm yourself with much patience, a clear head and, of course, the tools mentioned above.

The Models

The Compound of Two Cubes

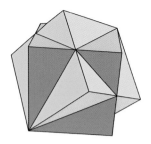

 Easy Intermediate

The first model in this collection is formed simply by two cubes that share a diagonal axis, around which the second cube is rotated by 60 degrees in relation to the first. At this specific rotation, the six equatorial edges intersect at their midpoints, giving rise to an aesthetically pleasing form. The two cubes should be made in contrasting colours to bring out the geometrical structure of the model, forming two 'carrousels' of alternating colours on the corners on the common diagonal axis. In Figure 1, the two cubes on the left come together to create the shape on the right.

Fig. 1 two cubes come together to form the carrousel

To make this model you will need a total of 13 square sheets of paper, six sheets each of two colours, plus one extra sheet to make a pre-folding template. I suggest using the largest square you can cut from an A4 or US Letter sheet (see page 13 for how to cut squares).

A pre-folding template is a folded piece of paper that helps you make certain creases on the module that would otherwise mark its faces. In this case, you will make a tool that allows you to make folds of both a third and a sixth of the width of the paper.

> **Top Tip: Clean hands, good lighting and an organised workspace all increase the quality and appearance of your finished model.**

Making the Pre-folding Template

The pre-folding template is used to make three preliminary creases in each of the 12 sheets before moving on to folding the module. Carefully follow the instructions on the following page to make the tool from a square sheet the same size as those used for the model. Try to be as accurate as possible in folding and be careful not to crumple the sheets as you work.

1

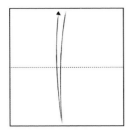

Make a crease along the horizontal centre line.

2

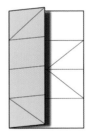

Bring the upper and lower edges to the centre line, crease and unfold.

3

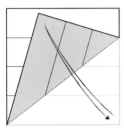

Crease from the bottom-left corner to the right end of the upper horizontal crease.

4

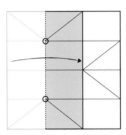

Crease from the top-left corner to the right end of the lower horizontal crease.

5

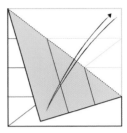

Fold over the left third of the model, using the reference points shown in the above diagram.

6

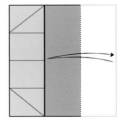

The finished template should look like this.

Using the Pre-folding Template

The following diagrams show you how to use the template to make the preliminary creases in the paper before moving on to fold the module.

1

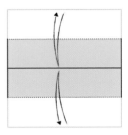

With the short part of the pre-folding tool facing up, next insert the module as far as it will go.

2

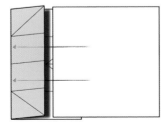

Bring the right edge of the module over to meet the right edge of the template and crease. The crease is at one third of the module.

3

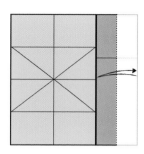

Turn the pre-folding tool over so that the long part is facing up, then rotate the module 90 degrees anti-clockwise and reinsert it.

4

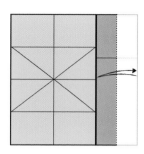

Bring the right edge of the module over to meet the right edge of the template and crease. This crease is at one sixth of the module.

5

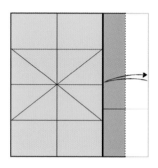

Rotate the module 180 degrees and repeat step 4 for the left edge.

6

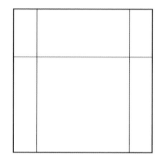

The module should now look like the above diagram.

Folding the Module

Once you have pre-folded all twelve sheets, you can now proceed to fold them into the completed modules that will go together to form the model.

Follow the instructions below very carefully, comparing your results with each diagram as you go. If there is some discrepancy, then go back and try to see what went wrong. Usually you can correct small errors without ruining the module, but if there are a few 'corrections' in a single module it might be better to make a completely new one. Don't trust yourself to remember every step of the folding sequence after a couple of modules. Instead, try to make each one better and more accurate than the last.

> **Top Tip:** Don't try to assemble the model as you make each module. The time they spend hanging on the unfinished model and the repeated build-unbuilding will wear and buckle the edges, giving your finished model a worn look. Apart from occasionally making a couple of test modules to see how the joints work, try to discipline yourself to make all the modules before starting the assemblage – the result will be a crisper, fresher model.

1

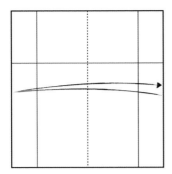

Crease along the vertical centre line.

2

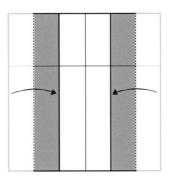

Fold both the outer sixths inwards.

3

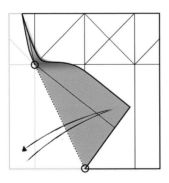

Fold the top corners inwards diagonally. Be careful to fold through all layers.

4

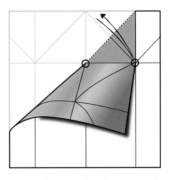

Unfold everything and make a top-right to bottom-left diagonal crease in the right one of the two squares at the top of the module.

5

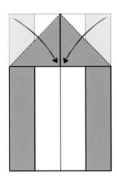

Crease from the bottom of the centre line up to the intersection of the horizontal crease and the vertical crease on the left.

6

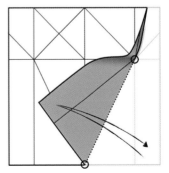

Crease from the bottom of the centre line up to the intersection of the horizontal crease and the right vertical crease.

7

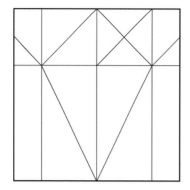

Check that the crease pattern looks the same as this.

8

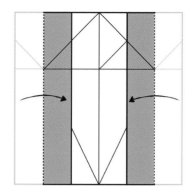

Fold the outer sixths inwards again.

9

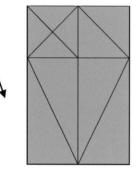

Flip the module over horizontally.

10

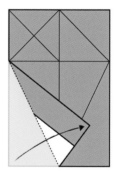

Fold the bottom-left corner along the diagonal made in step 6.

11

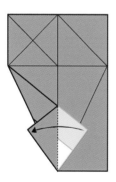

Fold the flap back so you can see the vertical centre line.

12

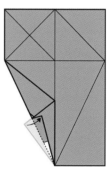

Fold the excess flap exactly back onto the module.

13

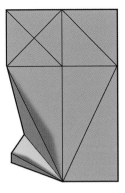

Gently unfold steps 10, 11 and 12.

14

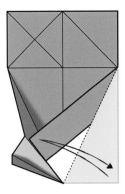

Crease along the diagonal made in step 5.

15

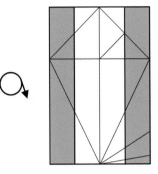

Flip the module back over horizontally.

16

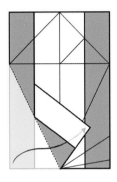

Make an inside reverse fold along the crease made in step 5.

17

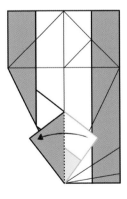

Fold the flap back so you can see the centre line.

18

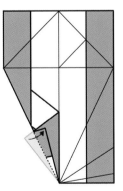

Fold the excess flap exactly back onto the module.

19

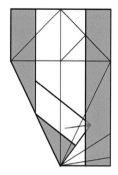

Gently unfold steps 17 and 18, tucking the tip of the flap under the folded right side sixth.

20

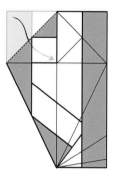

Make a reverse inside fold along the top left diagonal.

The following three diagrams are in '3D' view. The shading should help the folder understand what is happening.

21

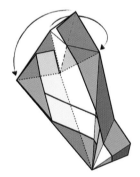

Fold along the vertical centre line and bring the top right of the module across to meet the left.

22

Fold the tab formed in step 20 over so that it protrudes out at a right angle.

23

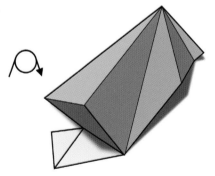

Flip the module over and check that it resembles the above diagram.

Assembling the Model

When you have folded all twelve modules, you can begin to assemble the model. Start by joining two modules of opposite colours at the ends with the protruding tabs. Make sure the tabs cross over each other as in Fig. 1. Slide the tabs gently into the slot on the other module until they meet.

Fig. 1 the internal joint of the protruding tabs (internal view)

Fig. 2 one completed pair (external view)

When you have joined the modules into six pairs, gently ease the long tabs into the pockets on modules of the opposite colour on another pair. Repeat this process for the other two modules on the same pairs. Gradually work round the whole model in this way. Take care not to squeeze the model as you put the final pieces into place and try not to damaging the 'carrousel' of narrow ends. Your finished model should look like the photograph below.

The Compound of Two Tetrahedra

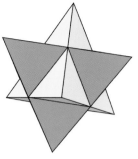

 Easy Intermediate

The Regular Tetrahedron is created by connecting the diagonals of the six faces of a cube. We say that the Tetrahedron is **inscribed** in the Cube (Fig. 1).

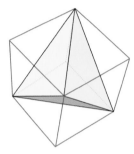

Fig. 1 the Tetrahedron inscribed in a cube (red frame)

It should be clear from figure 1 that only one of the two diagonals has been used on each face. It is also possible to create a tetrahedron using the other diagonal on each face. If both tetrahedra are generated within the same cube, the result is the Compound of Two Tetrahedra (Fig. 2).

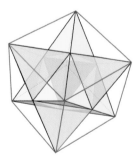

Fig. 2 two tetrahedra inscribed within the same cube

To make this model you will need 24 square sheets of paper, 12 each of two colours. Use quite stiffish paper and make crisp accurate folds. Use fairly large squares so you will be able to get your fingers inside towards the end of the assembly process. Follow the instructions below carefully and check your results against the diagrams.

Folding the Module

1

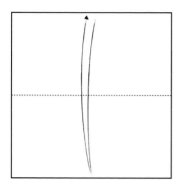

Crease the horizontal centre line.

2

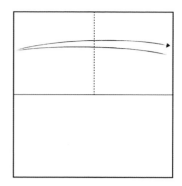

Crease the vertical centre line only from the top edge to the horizontal centre line.

3

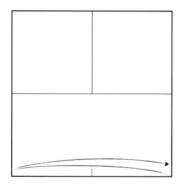

Make α reference crease at the half-way points of the bottom edge.

4

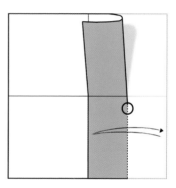

Using the reference creases for alignment, make a vertical crease one quarter from the right edge that stops short of the centre line.

5

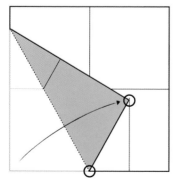

Bring the bottom-right corner over so that the bottom edge lies on the fold made in step 4 and fold.

6

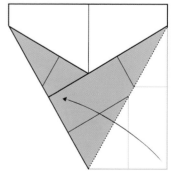

Bring the bottom-right corner over so that the bottom edge lies on the fold made in step 5 and fold.

7

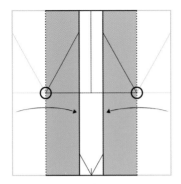

Using the intersections as guides, fold the two outer vertical portions towards the centre.

8

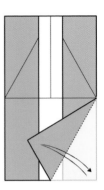

From here on, the reference creases have been removed. Fold and unfold the bottom-right corner along the existing crease on the reverse side made in step 6.

9

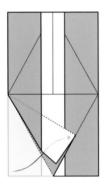

Make an inside reverse fold on the bottom-left corner as shown in the diagram, then unfold everything.

10

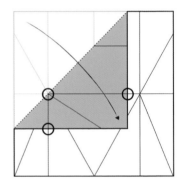

Fold the top-left corner, by fixing the intersection of the horizontal centre lines and the left vertical crease. Align the reference points before creasing.

11

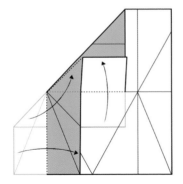

Make a reverse inside fold along the diagonal fold made in step 10 and the left vertical crease.

12

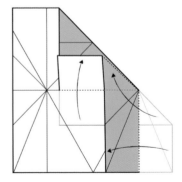

Repeat steps 10 and 11 for the right side of the module, then unfold everything.

13

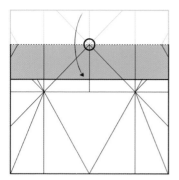

Fold the top edge of the module along the horizontal line that crosses the intersection of the diagonal creases made in steps 10 to 12.

14

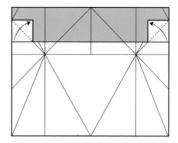

Fold the corners of the fold made on step 13 back up diagonally along the 45 degree creases made in steps 17 and 18.

15

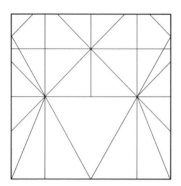

Check that the crease pattern matches the above diagram. Remember that the reference creases have been removed.

16

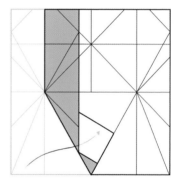

Refold the reverse inside fold in step 9.

17

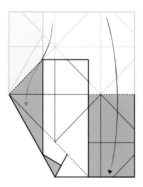

Bring the top half of the module down and make a reverse inside fold along the 45 degree diagonal at the upper-left.

18

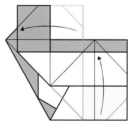

Swivel the top flap made in step 17 and the upper layer of the bottom half of the module anti-clockwise.

19

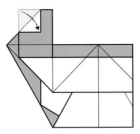

Refold the left flap made in step 14.

20

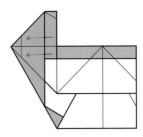

Tuck the flap on the right completely inside the pocket on the extreme left.

21

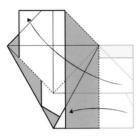

Swivel fold the top layer on the right of the module clockwise inverting the 45 degree diagonal to reach the diagram. This might take a few shots to get right.

22

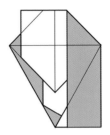

After these two quite odd maneuvers, check that your module resembles the above diagram before moving on.

23

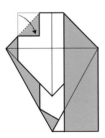

Fold down the top-left corner.

24

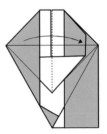

Fold the flap on the left upper layer over on to the right side of the module. Press the module flat.

25

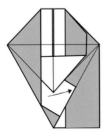

Bring the end of the innermost fold out so that it is in front of the right hand vertical fold.

26

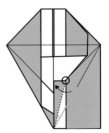

Fold back the flap brought out in step 25 along the line from the bottom-left tip to the point where it emerges (marked with a circle).

27

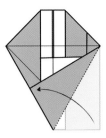

Fold the bottom-right corner over so that the bottom edge lies on the left diagonal edge.

28

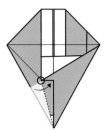

Fold the edge of the flap made in step 27 from the bottom apex to the point where it crosses the inside edge of the left vertical fold.

29

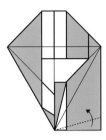

Unfold steps 27-28 and reverse the fold made in step 28.

30

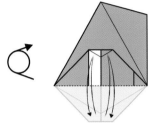

Flip the module over vertically. Fold and unfold the two tabs, reinforcing the crease along horizontal centre line.

31

The finished module should resemble the above diagram.

Assembling the Model

This model uses an ingenious split tab-slot arrangement that makes a puzzle-like joint between the modules of different colours. Place two modules of different colours outer face-down with the double tabs opposite each other. Put them together with the 'I' inside the 'V' shapes on both sides (Fig. 3a).

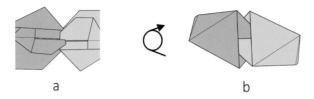

a   b

Fig. 3 beginning to assemble the split tab joint

Gently push the two modules together to reach a point where they won't go any farther and stop. Make sure the tabs on the outside of the model are on the inside (Fig. 3b). Now slide the two modules laterally apart as shown in figure 4a, so that the edges of the tabs can be seen.

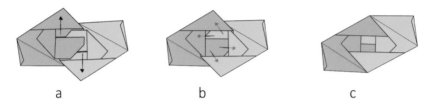

a   b   c

Fig. 4 using the internal pockets to finish the joint (other details removed)

Slide the modules back together, tucking the two tabs inside all of their respective internal pockets as you go (Fig. 4b). Finally, push the two modules completely together (Fig. 4c). Make 12 pairs of modules like this.

The joint between the remaining edges is made in two steps. Firstly, the tab is inserted into the slot in the usual manner (Fig. 5a). Then the internal tab is turned back under the large right-angled flap in the middle of the module (Fig. 5b).

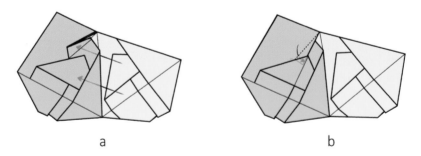

a   b

Fig. 5 the joint between the remaining edges

Connect four pairs to make the base of the model, then add the vertical sides and finally the top four pairs. The final couple of joints should stay in place even without the inner tuck in Fig. 5b, so don't worry.

The Compound of an Octahedron and a Cube

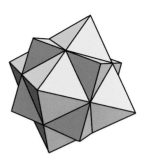

 Intermediate

This form belongs to a group of polyhedra known as **dual compounds**. The term **dual** is used to describe a polyhedron derived by a specific geometrical process, which involves taking the midpoint of every edge of a polyhedron and making a line that extends perpendicular to that edge at an equal angle from the adjoining faces. For example, figure 1a shows an edge formed by two squares joined at right angles. A line extends from the midpoint of the edge, and it is also raised by 45 degrees from the plane of either square.

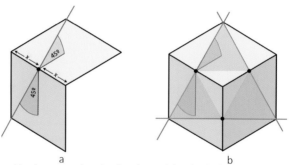

Fig. 1 generating the first face of the dual of a cube

By adding another square at right angles to the other two, and then performing the same process on the two newly formed edges, a polygon is generated – in this case it is an equilateral triangle (Fig. 1b). Note that if there were more planes converging at the vertex, the generated polygon would have correspondingly more sides. When this process is applied to all the edges of a polyhedron, a new one is formed, and that polyhedron is known as the **dual** of the first. You can see from the picture at the top of the page that each vertex of the cube has generated an equilateral triangle and together these form an Octahedron, thus **the Octahedron is the dual of the Cube**. Performing the same process with an octahedron will generate a cube and since there are four converging faces at each vertex of the Octahedron, each generated polygon will have four sides – in this case, a square. All dual pairs have this two-way quality, and their compounds are quite beautiful.

To make this model you will need 48 square sheets, 24 each in two different colours. Use paper that is resilient since there are some tough moves in the assembly stage.

There are two modules, one for each colour. For this model, folding accurately is important since the joints are tight fits with a lot of layers. There is a lot of work here, but the solidity of the finished model is worth it. Follow the instructions and check your results with the diagrams.

> **Top Tip: To avoid eye strain, take regular breaks. While working, look up at distant objects so the muscles in your eye lenses get some rest, too. Origami is not a race, so if you feel tired, just stop and have a break.**

Folding Module A – the Cube (make 24 in one colour)

1

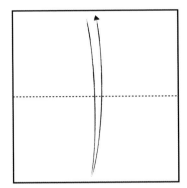

Make a crease along the horizontal centre line.

2

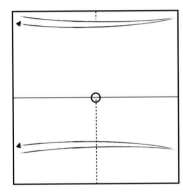

Crease the vertical centre line from the bottom edge as far as the horizontal centre line and mark the midpoint of the top edge.

3

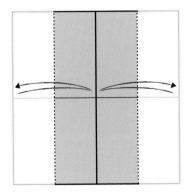

Fold the left and right edges in to the vertical centre line and unfold.

4

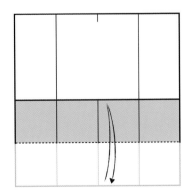

Fold the bottom edge up to the horizontal centre line and unfold.

5

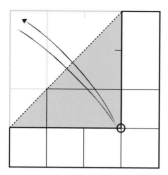

Fold the top-left corner diagonally to meet the intersection of the lower and right quarter lines, then unfold. Try not to go beyond the creases.

6

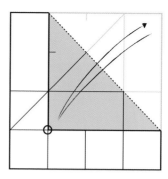

Fold the top-right corner diagonally to meet the intersection of the lower and left quarter lines, then unfold. Again, try not to go beyond the creases.

7

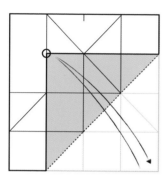

Fold the bottom-right corner diagonally to meet the intersection of the upper and left quarter lines, then unfold. Again, try not to go beyond the creases.

8

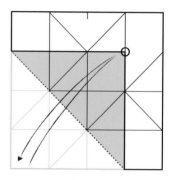

Fold the bottom-left corner diagonally to meet the intersection of the upper and right quarter lines, then unfold. Try not to go beyond the creases.

9

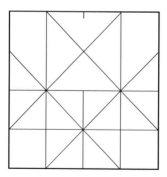

Before continuing, check that your crease pattern resembles the above diagram.

10

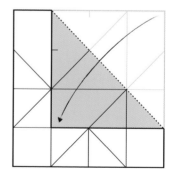

Refold step 6.

11

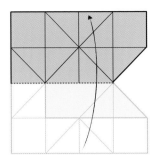

Fold the entire bottom half of the module up on to the top half.

12

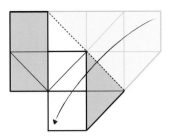

Fold the right part of the module over along the diagonal running from the top end of the left quarter line.

13

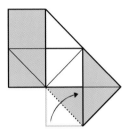

Fold the square tab at the bottom of the module upwards diagonally.

14

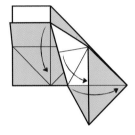

This and the following diagram show the stages of a swivel fold, bringing the top layer down to meet the centre line.

15

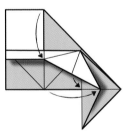

The whole of the top later of the middle third of the module swings across to the right until the extreme-right tips line up.

16

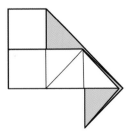

When the swivel fold is complete, the module should look like the above diagram.

17

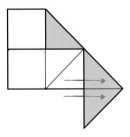

Roll back the top layer of the extreme right and insert it into the pocket on the bottom layer.

18

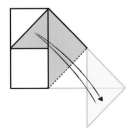

Gently fold and unfold the lower-right half of the module up diagonally.

19

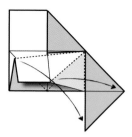

A second swivel fold begins by pulling the top layer at the horizontal centre line downwards and over to the right.

20

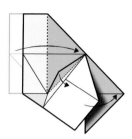

Fold the left quarter of the module over, while bringing the top layer over so that it completely covers the right part of the module.

21

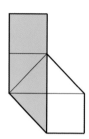

When the second swivel fold is complete, the module should look like the above diagram.

22

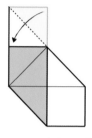

Fold the top square tab down diagonally.

23

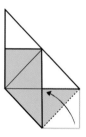

Fold the bottom square tab up diagonally.

24

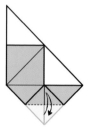

Bring the bottom tip of the module up to meet the horizontal centre line and crease.

25

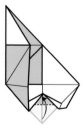

Refold step 24, inverting the folds so the bottom tip is now on the inside.

26

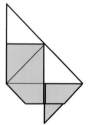

Your module should now look like the above diagram.

27

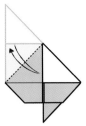

Gently reinforce the diagonal crease extending up from the left edge of the horizontal centre line.

28

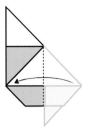

Fold the top layer of the right side over to the left.

29

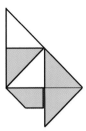

Your module should now look like the above diagram.

30

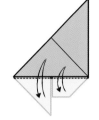

Flip the module over horizontally and gently crease the two extruding tabs. Don't force these too much, since they might damage the face.

31

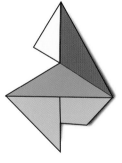

Gently ease the module out so that it resembles the above diagram.

Folding Module B – the Octahedron (make 24)

The second module has a lot in common with the first. The two tabs at the bottom of Module A are reproduced in Module B, but the upper half of the module is now going constitute part of the Octahedral face and, therefore, an equilateral triangle replaces the right-angled triangle of module A. The first few folds are the same but soon Module B takes its own independent road. Take your time to make the folds as accurately as you can since the assembly will involve some tight fits and folding through several layers. As above, hints on how to make sure these are manageable are included in the captions.

1

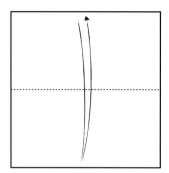

Make a crease along the horizontal centre line.

2

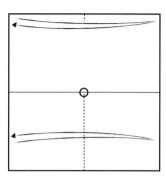

Crease the vertical centre line from the bottom edge as far as the horizontal centre line and mark the midpoint of the top edge.

3

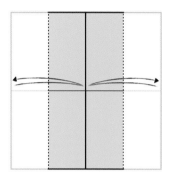

Fold the left and right edges in to the vertical centre line and unfold.

4

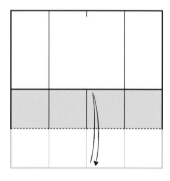

Fold the bottom edge up to the horizontal centre line and unfold.

5

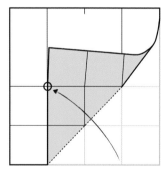

Fold diagonally from the bottom of the left vertical crease to the intersection of the horizontal centre line and the right vertical crease. Align by using the reference point indicated in the diagram.

6

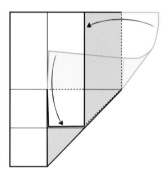

Fold the top of the flap made in step 5 down along the horizontal centre line, and fold the right quarter inwards.

7

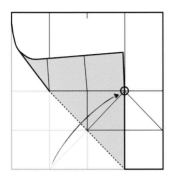

Repeat step 5 for the other bottom corner.

8

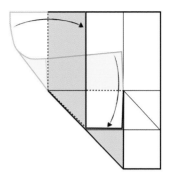

Repeat step 6 for the other side.

9

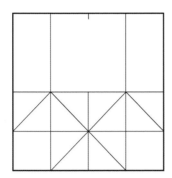

Unfold and compare your crease pattern with the above diagram.

10

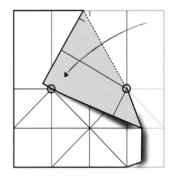

Fixing the intersection of the Horizontal centre line and the right vertical crease, fold the top-right corner over so that the right edge falls on the intersection of the Horizontal centre line and the left vertical crease.

11

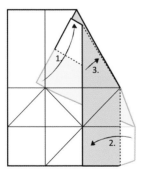

Make an inside reverse fold on the flap made in step 10.

12

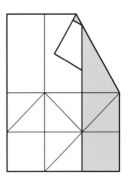

The completed fold should look like the above diagram.

13

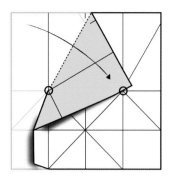

Unfold and repeat step 10 for the left side.

14

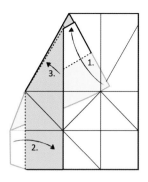

Repeat step 11 for the left side.

15

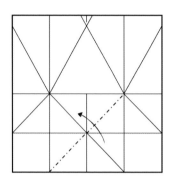

Unfold the module and make a mountain crease along the diagonal running from the bottom end of the left vertical crease to the intersection of the right vertical crease and the horizontal centre line.

16

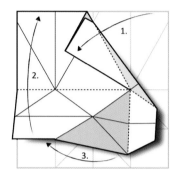

Refold the top half of step 12 (1). Bring the bottom half upwards (2). Fold along the mountain crease made in step 15 (3).

17

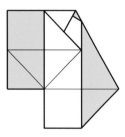

When the maneuver started in step 16 is completed the module should resemble the above diagram.

18

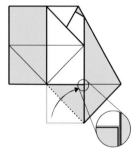

Fold up the loose corner at the bottom of the module. Make sure you stay inside the creases on the bottom layer.

19

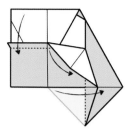

Make a swivel fold, bringing the top of the top layer down to the horizontal centre line, while pushing the lower left portion over to the right..

20

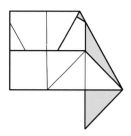

When step 19 is completed, the module should resemble the above diagram.

21

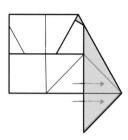

Roll back the top layer of the extreme right and insert it into the pocket on the bottom layer.

22

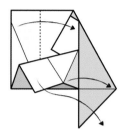

A second swivel fold begins by pulling the top layer at the horizontal centre line downwards and over to the right.

23

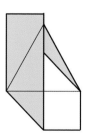

When the second swivel fold is complete, the module should look like the above diagram.

24

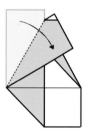

Fold the top left corner over along the existing diagonal.

25

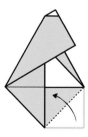

Fold the bottom-right corner up diagonally.

26

Bring the bottom tip of the module up to meet the horizontal centre line and crease.

27

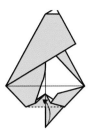

Refold step 26, inverting the folds so that the bottom tip is now on the inside.

28

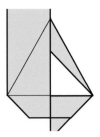

Unfold the top-left corner. The module should now resemble the above diagram.

29

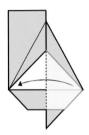

Fold the top layer of the right side over to the left.

30

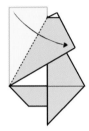

Now refold step 24.

31

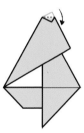

Fold over the extreme upper tip onto the module.

32

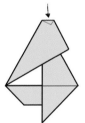

Tuck the tip inside.

33

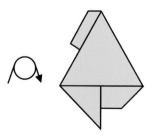

Flip the module over horizontally.

34

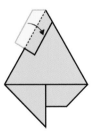

Fold the extruding top-left flap over onto the module.

35

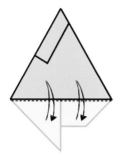

Gently crease the two extruding tabs.
Don't force these too much, since they
might damage the face.

36

Gently ease the module out so that it
resembles the above diagram.

Assembling the Model

Begin by making the corners of the cube. This is quite tough since, at first, the modules do not seem to fit together. Only when they are gradually brought into their three-dimensional position do they start to slide into place. Take three A modules and place them as in figure 2a, then gradually push them inwards to form one corner of the cube (Fig. 2b). To finish the joints, turn the three modules over and massage them, two at a time, into place (Fig. 2c).

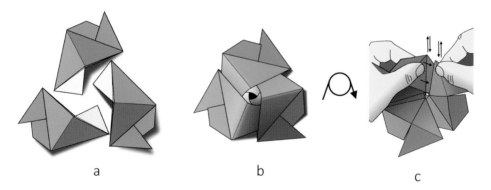

a   b   c

Fig. 2 building the cubic corner

The next step is to add the three B modules that make up one face of the octahedron to each of the eight cubic corners. For the tab and slot on each B module, there is a corresponding slot and tab on the A module. Begin by simultaneously inserting the point of the tabs into the extreme corner of the slots (Fig. 3a). Make sure that the 'square' end of the tab on the A module is towards the outside of the model. Next, simultaneously slide the square side of the tabs into the slots so that the two modules come together (Fig. 3b).

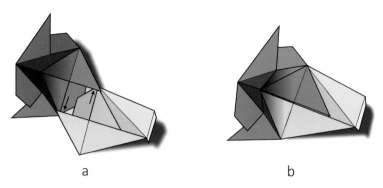

a	b

Fig. 3 adding the octahedral modules

When you have created all eight of the octahedral faces then join them together by inserting the tabs into the corresponding slots on the B modules (Fig.4). Continue connecting the sides until you have built the complete octahedron, then push any lose joints firmly into place.

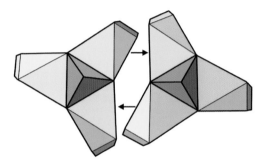

Fig. 4 joining the octahedral faces together

The finished model should be very sturdy and, if you are using these models in a classroom situation, this one can be passed around without much fear of it collapsing even in the clumsiest hands.

The Compound of Three Cubes

 Intermediate/Advanced

The Compound of Three Cubes appears in the famous book *Polygons* and *Polyhedra, Theory and History* (1901) by German mathematician Max Bruckner (1860–1934). Bruckner created models of dozens of the polyhedra described in his book and presented them in a series of photographic plates. Number 23 on Plate IX shows the Compound of Three Cubes. This form also appears in Mauritz Cornelius Escher's lithograph *Waterval* (1961) atop one of the columns supporting a paradoxical waterfall.

Geometrically, the form comprises three cubes with a common axis that passes through opposing faces of each cube at the point (P) where the line connecting the point C (where the arc, radius AB, bisects the lower edge) and D the midpoint of the adjacent edge (line CD) intersect with the vertical centre line (Fig. 1).

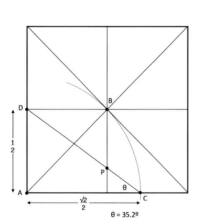

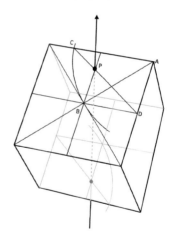

Fig. 1 Defining the entry and exit points for the common axis

Each of the two subsequent cubes is the result of rotating the original cube in Fig. 1 about the common axis by 120 and 240 degrees respectively.

The desired angle of 35.2 degrees (θ in Fig. 1) is reproduced in the model with less than a 0.2% error (steps 11-14 in Unit B). This tiny discrepancy is indiscernible in the finished model, even if built on a large scale.

Preparing the Paper

This model requires 48 square sheets of paper cut from 18 sheets of A4, six each in three colours. The first 24 sheets can be cut from twelve sheets of A4, dividing the long side in half (Fig 2a). The second 24 sheets can be cut from six sheets of A4 dividing the short side in half (fig. 2b).

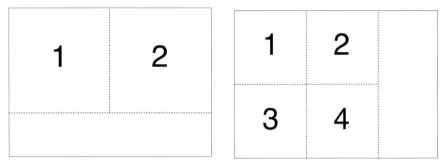

Fig. 2a Unit A   Fig. 2b Unit B

The larger squares are folded into the Unit A modules and the smaller squares are folded into the Unit B modules. The ratio between the edge length of the two sizes is $1:\sqrt{2}$, the same ration as that between the edge and the diagonal of a square. Luckily for us, A4 paper is designed exactly in these proportions (Fig 3).

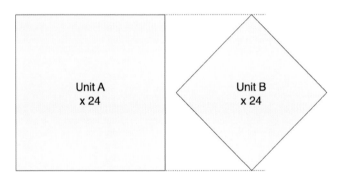

Fig. 3 the proportions of the Unit A and Unit B modules

Top Tip: Use stiffish 80 g/m² paper for best results, and chose three highly contrasting colours to make the cubes of the finished model stand out.

Folding the Unit A Module

Make 24 Unit A modules, eight in each of the three colours. Once again, follow the instructions below very carefully, comparing your results with the diagrams as you go. Try to make every module better and more accurate than the last.

1

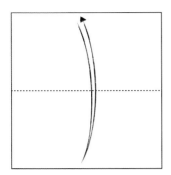

Make a crease along the horizontal centre line.

2

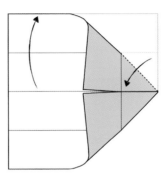

Fold and unfold the lower half. Fold the top edge down to the centre line.

3

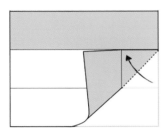

Bring the bottom-right corner up towards the centre line and fold only half of the diagonal.

4

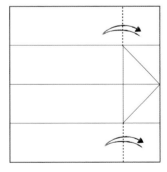

Unfold step 2 and repeat step 3 for the top-right corner. Again, take care to fold only half of the diagonal.

5

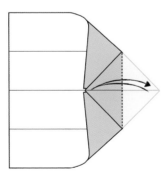

Holding the two corners in place from steps 3 and 4, fold and unfold the extreme right corner into the centre.

6

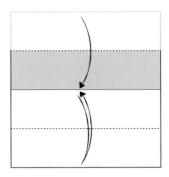

Unfold the paper and extend the vertical fold from step 5 to the upper and lower edges.

7

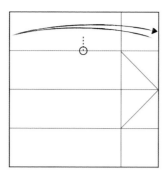

Bring the top-left and right corners together and mark the half way only on the top side of the fold from step 2.

8

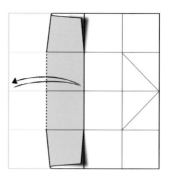

Using the half way mark, fold the left quarter over and crease only the middle two quarters.

9

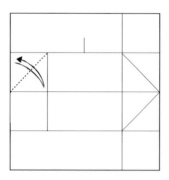

Fold the diagonal of the second square in the far-left column from the centre line inwards.

10

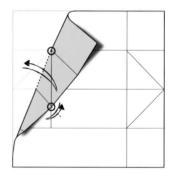

Using the top of the crease made in step 8 as a reference point, bring the left end of the horizontal centre line onto the left vertical fold and make a reference crease at the intersection.

11

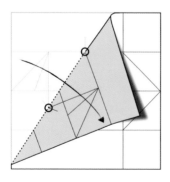

Fold from the reference crease made in step 10 to the half-way point marked in step 7, then extend the crease to the left edge, but not to the top edge.

12

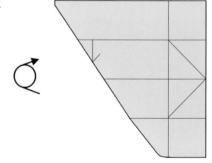

Flip the model over vertically. This begins the 'mirroring' technique described in the Tools and Techniques section on page XX.

13

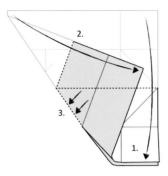

1. Bring the top-right corner down to reveal the fold made in step 11.

2. Bring the top-left corner over diagonally.

3. Press out only half of the new fold.

14

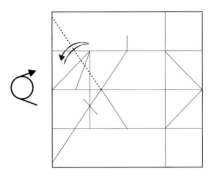

Flip the module over vertically and unfold everything. Extend the crease made in step 13 all the way to the left edge. This completes the mirror.

15

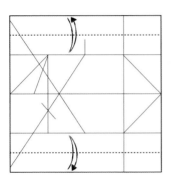

Crease the top and bottom quarters into eighths. These creases will be reversed later.

16

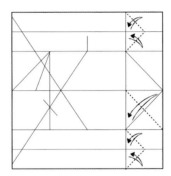

One by one, make the five diagonal creases shown in the above diagram. These will help fold the 'gable end' of the module.

17

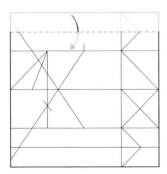

Fold the top eighth of the module to the back, making a mountain fold.

18

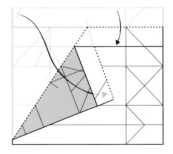

Make a reverse inside fold about the mid-point of the top-quarter horizontal line.

19

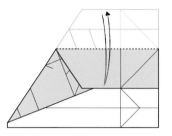

Crease along the horizontal centre line, folding through all the layers.

20

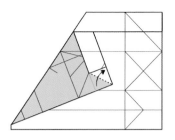

Fold back the tip of the flap made in step 19, so that only one layer of paper lies on the three-quarters horizontal line.

21

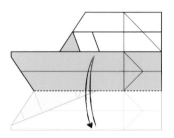

Crease along the three-quarter horizontal line, folding through all layers.

22

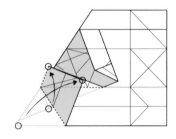

Bring the lower-left corner up, so that the edges are parallel at the points marked.

23

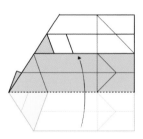

Fold along the three-quarter horizontal line.

24

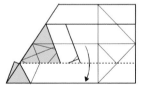

Fold the lowest eighth of the module back down to the bottom edge.

25

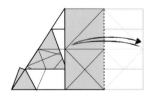

Reinforce the vertical fold.

26

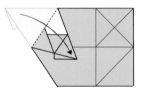

Flip the module over horizontally. Fold the top left corner over along the crease made in step 11.

27

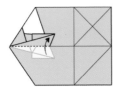

Fold the tip of the flap made in step 26 along the centre line.

28

 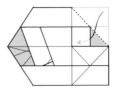

Flip the model over horizontally and tuck the top-right corner under the flap created by the top quarter fold.

29

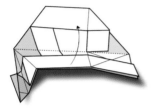

(This and the following three diagrams are in 3D view.) Fold gently along the centre line.

30

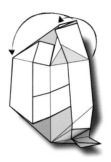

Rotate the module 90 degrees anti-clockwise. Fold along the long crease made in step 16 to bring the top-right corner inside.

31

Finally, push the extruding top tab over the edge and fold. This completes the Unit A module.

32

Flip the completed module over and ease out the tabs. It should resemble the above diagram.

When you have finished all 24 Unit A modules, you can then move onto the Unit B modules. Again there are 24, eight each in three colours. Remember to use the smaller square for the Unit B module.

Once again, follow the instructions on the following pages very carefully, comparing your results with each diagram as you go, and try to make every module better and more accurate than the last. By the way, this is one modules that I really enjoy making because some of the folds are very satisfying, such as when the outer edges magically line up in step 6.

Folding the Unit B Module

1

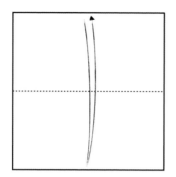

Crease along the horizontal centre line.

2

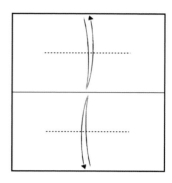

Crease the upper and lower halves, but not all the way to the edges.

3

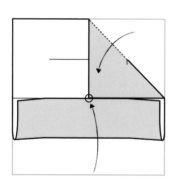

Bring the top right corner in to the middle and crease only the top half.

4

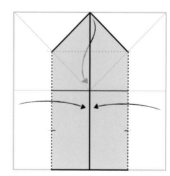

Repeat step 3 for the top left corner, now aligning it with the right corner.

5

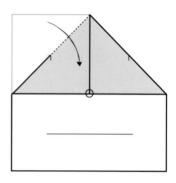

Unfold the module and make two diagonal folds that link the intersections of the previous folds with the top corners.

6

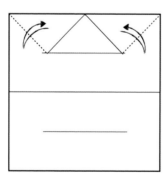

Fold the midpoint of the top edge down to the middle, then fold the left and right edges to the centre. Be sure to align the edges before creasing.

7

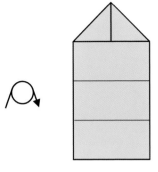

Flip the module over horizontally.

8

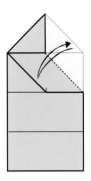

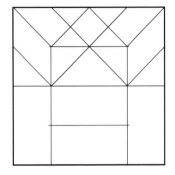

Diagonally fold and unfold the top right 'wing' so that it lies exactly on the horizontal centre line.

9

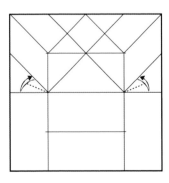

Repeat step 8 for the top left wing.

10

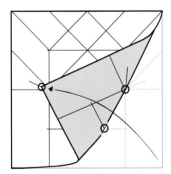

Flip the module over horizontally and open out all the folds. Check that your module resembles the above diagram. You may sharpen some of the creases.

11

Make two small reference creases by folding the diagonal creases made in steps 8 and 9 down to meet the horizontal centre line.

12

Fix the intersection of the centre line and right vertical crease, bring the bottom right corner to the reference crease made in step 11, and crease the point where the fold crosses the lower horizontal crease.

13

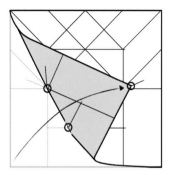

Repeat step 12 for the left side of the module.

14

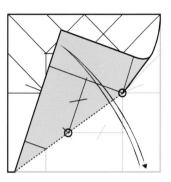

Crease the line connecting the mark made in step 13 and intersection of the centre line and right vertical crease. Extend it to the bottom edge.

15

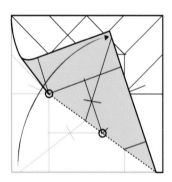

Fold the line connecting the mark made in step 12 and and intersection of the centre line and left vertical crease. Extend it to the bottom edge.

16

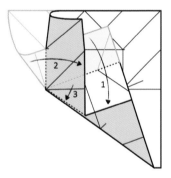

Make an inside reverse fold, bringing the tip of the flap made in step 15 down and the left edge over it. Fold only the lower half of the left vertical fold.

17

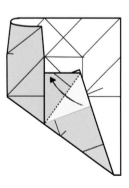

Fold back the tip of the flap made in step 16 so that its left edge is just touching the left edge of the module.

18

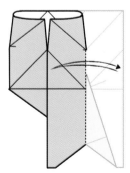

Without creasing the top quarter, fold and unfold the right quarter of the module, through all the layers.

19

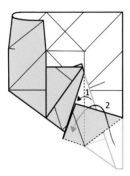

1. Fold the top layer of the right part of the module along the crease made in step 18.
2. Bring the lower-right corner up parallel to the top layer and fold.

20

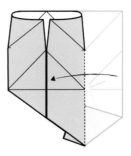

Bring the right quarter of the module over. Do not fold the top quarter but fold the remainder of the crease through all the layers.

21

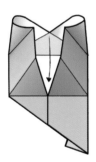

Bring the mid-point of the top edge internally into the centre, recreating the fold in step 6.

22

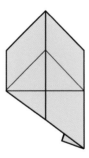

After step 21, the module should now resemble the above diagram.

23

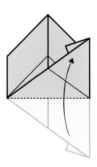

Fold the bottom half of the module up along the horizontal centre line.

24

Fold the tab made in step 23 back down along the existing diagonal crease.

25

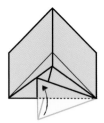
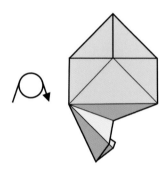

Bring the tip of the flap made in step 24 up along the horizontal centre line, again folding through all layers.

26

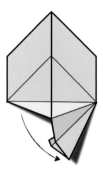

Pull out the folds made in steps 23-25 not so much that they sit flat.

27

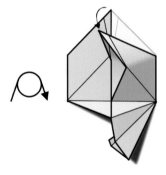

Flip the module over horizontally.

28

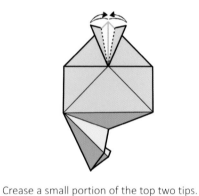

Crease a small portion of the top two tips. Judge this by making sure the tip points slightly upwards.

29

Flip the module over horizontally and push the right tip down so that it opens up.

30

Carefully fold all of the outside part of the crease made in step 28 inside the opening.

31

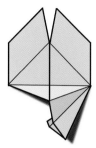

Repeat steps 29 and 30 for the left tip and press the top half of the module flat.

32

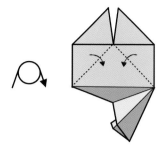

Flip the module over horizontally and reinforce the creases of the two wings made in steps 8 and 9.

33

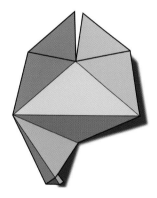

The finished Unit B module should now resemble the above diagram.

Assembling the Model

After making all 48 modules, you should begin assembly by making the eight three-part submodules shown below. You should make four clockwise (Fig. 4a) and four anti-clockwise (Fig. 4b) versions, using all three colours of the Unit B modules.

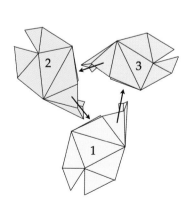

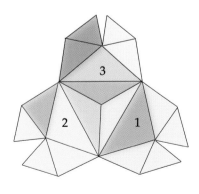

a. clockwise submodule assembly (make 4)

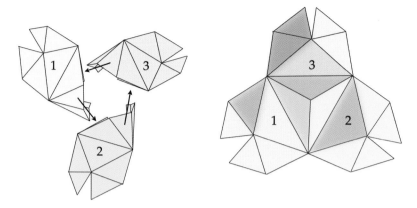

b. anticlockwise submodule assembly (make 4)

Fig. 4 assembly of the three-part submodules

Next you should make two 'crosses' with two colours, using the Unit A modules as shown in figure 5. Note that the double folds of the joints interlace. These crosses will become the top and the base of the model.

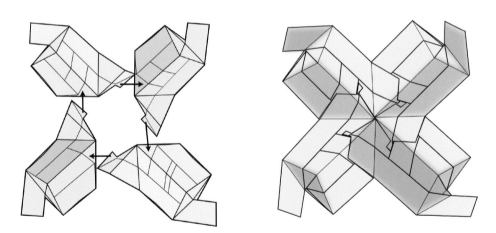

Fig. 5 assembly of the base cross

Lay the first cross face down on your assembly board. Insert the tab with the third colour on the three-sided modules into the right-angled gap of the first cross. Join the third coloured Unit A to the first cross using the double tab and slot arrangement. Build up the model until you reach the point where you add the final cross. Then work perimetrically round the joints, easing them as need be to get the final part into place.

The Compound of Five Octahedra

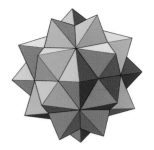

 Intermediate

This form, too, appears in the plates of Max Brückner's *Polygons* and *Polyhedra, Theory and History,* published in 1901 (Number 6 on Plate IX), and has been the subject of several origami models over the past century. This version is probably neither better nor worse than any of the others, but the internal design is original, having an approach to the assembly procedure that uses the triangular rather than the pentagonal base.

The model consists of a regular octahedra rotated by 31.55 degrees about the z axis and then repeatedly rotated about the y-axis at increments of 72 degrees. Ideally, the faces should have angles of 60 degrees, 44.5 degrees, and 74.5 degrees but this design rounds these last two to 45 degrees, and 75 degrees (Fig. 1).

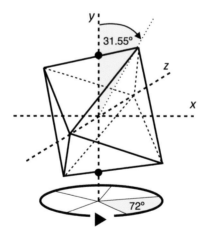

Fig. 1 the axis and rotation of the five octahedra

You will need 30 square sheets of paper, six each of five colours. They should not be much smaller than 10.5 × 10.5 cm since there is a tricky maneuver at step 34 of the module, which would be rather fiddly with smaller paper. Use a stiff 80 g/m² paper and crease all the folds well. This model also looks good with five colours that are somehow related, such as a series of blue hues (Indigo, Navy blue, Sky blue, Turquoise, pale blue) as well as the usual high-contrast sets.

Folding the Module

1

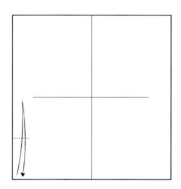

Crease the vertical centre line.

2

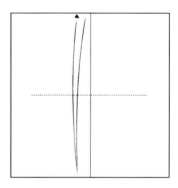

Crease the horizontal centre line but not all the way to the edges.

3

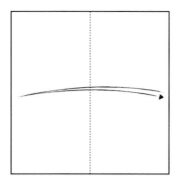

Make a short horizontal reference crease a quarter of the way up the left edge.

4

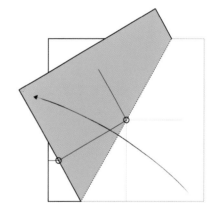

Keeping the centre fixed, fold the bottom-right corner so that the vertical centre line meets the reference crease made in step 3.

5

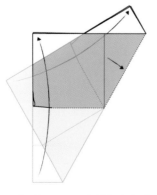

Make an inside reverse fold by bringing the bottom half upwards and folding over the vertical centre line. Then press out the resulting fold.

6

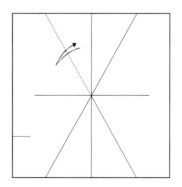

Unfold everything, then extend the crease made in step 5 all the way to the top edge.

7

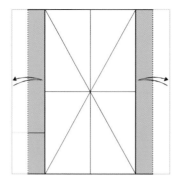

Fold the left and right sides over so that the corners meet the ends of the diagonal creases.

8

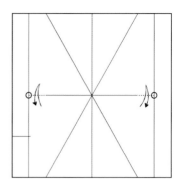

Extend the ends of the horizontal crease made in step 2 so that they meet the creases made in step 7.

9

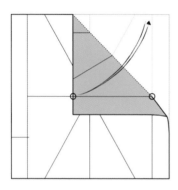

Make a diagonal crease by bringing the right vertical crease over to meet the centre line, while keeping the shown intersection fixed. Do not fold the portion to the right.

10

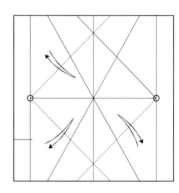

Use the horizontal and vertical centre line creases to mirror the crease made in step 9 on the remaining three quarters.

11

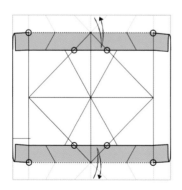

Make horizontal creases at the intersections of the creases made in steps 9 and 10, by aligning the diagonals and the centre line. Again, do not fold beyond the left and right vertical creases.

12

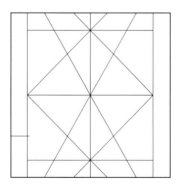

The crease pattern should now resemble the above diagram.

13

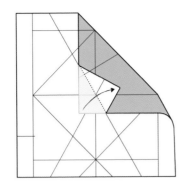

Refold step 9, then fold back the portion of the flap to the left of the top-left bottom-right diagonal.

14

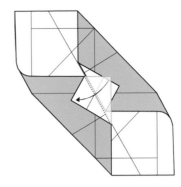

Do the same for the bottom-left corner.

15

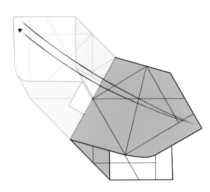

Fold the module along the top-right bottom-left diagonal, taking in all the layers, then unfold.

16

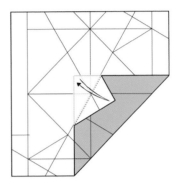

Bring the bottom-right corner up along the 45 degree diagonal and crease the tip along the top-right bottom-left diagonal.

17

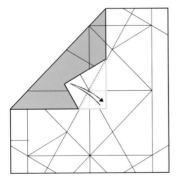

Bring the top-left corner down along the 45 degree diagonal and crease the tip along the top-right bottom-left diagonal.

18

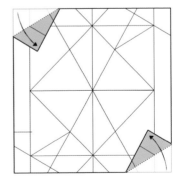

Unfold the module and refold the extreme corners creased in steps 16 and 17 inwards.

19

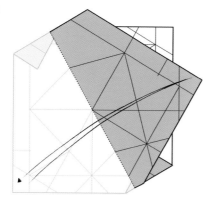

Crease the entire module along the top-left bottom-right diagonal taking in the layers at the folded corners.

20

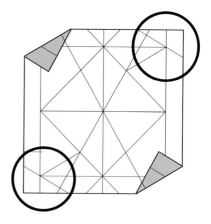

In the next two steps you will make a short crease perpendicular to the creases made in steps 13 and 14.

21

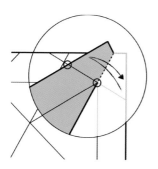

Fixing the intersection of the crease made in step 13 and the right vertical crease, make a perpendicular crease from there up to the top edge.

22

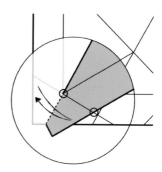

Fixing the intersection of the crease made in step 14 and the left vertical crease, make a perpendicular crease from there down to the bottom edge.

23

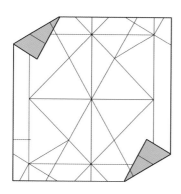

Your module should now look like the above diagram.

24

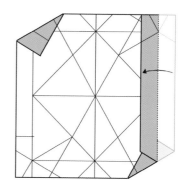

Fold over the extreme right portion of the module, taking in the layer at the bottom corner.

25

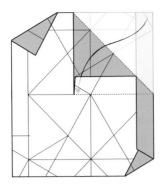

Make an inside reverse fold along the top right 45 degree diagonal.

26

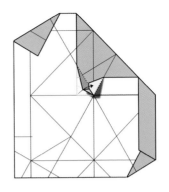

(3D image) Gradually bring the tip of the fold made in step 25 upwards so that it stands up pointing out of the folding plane.

27

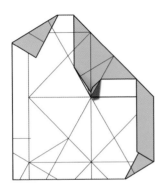

The module should now look like the above diagram.

28

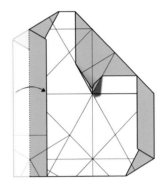

Fold over the extreme left portion of the module, taking in the layer at the top corner.

29 **30**

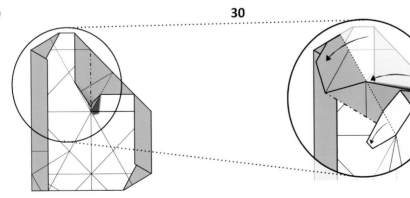

Make the preliminary folds as shown above and start to push the top-right side of the module over.

The inside (mountain) fold should be on the inside of the tip brought out in step 26-27. Make sure that the module resembles the above diagram.

31

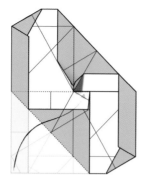

Unfold step 30, and make an inside reverse fold along the bottom-left 45 degree diagonal.

32

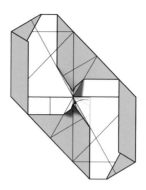

Once again, bring the tip up so that it points out of the folding plane.

33

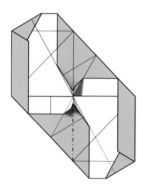

Make the preliminary mountain fold and valley folds as shown above. Start to bring the left half of the module over to the right.

34

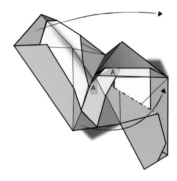

When the tip brought up in step 32 comes over, tuck it behind the fold made in step 25. Then slide the tip shown in step 27 behind the fold made in step 31. Now push the module closed.

35

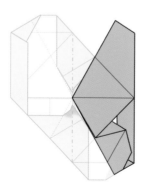

With that maneuver completed, the module should resemble the above diagram. Take some time here to flatten the model smoothly.

36

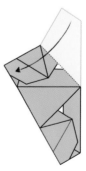

Fold the top-right section down along the existing diagonal.

37

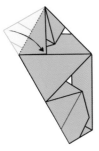

Fold the tip back over onto the module.

38

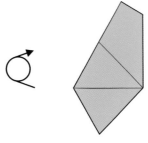

Flip the module over vertically.

39

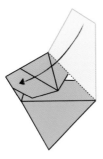

Fold the top-right section down along the existing diagonal.

40

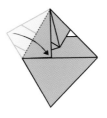

Again, fold the tip back over onto the module.

41

Pull the module gently open and smarten up the edges. It should now look like the above diagram.

Assembling the Model

All the units join together in the same way. The internal flaps interlock, keeping the units in place as shown in Figure 2.

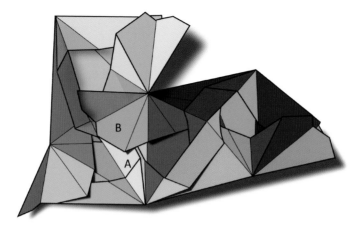

Fig. 2 the internal joint between the units.

The internal flap marked 'A' flips down, allowing the other module to fit in place with flap 'B' raised. Then flap A flips back into place and finally flap B closes on top to complete the joint. You should try to find this mechanism and understand it before embarking on the full assembly. Take a couple of units and try it out. Now, join a third unit of a different colours. Make two more submodules of three units using the colour scheme in Figure 3 and join them all together.

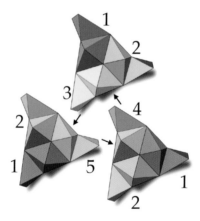

Fig. 3 the colour scheme

Fill the gaps with three more units with colours 3, 4 and 5 in opposing corners. Then work upwards following the same approach to the colour scheme. When finished, the model should be quite sturdy.

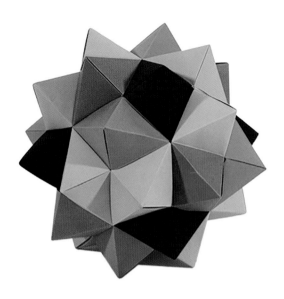

The Compound of
Three Octahedra

 Advanced

This form also appears in the photographic plates of Max Brückner's *Polygons and Polyhedra, Theory and History* (1901) as Number 12 on Plate VIII and, as with the Compound of Three Cubes, it was used by M. C. Esher, this time in his woodcut *Stars* (1948) where you can also see the compound of Two Cubes floating off to the left.

The form consist of one regular octahedron rotated by 29.36 degrees about the z axis, which is then rotated twice about the y-axis at increments of 120 degrees to generate the other two (Fig. 1).

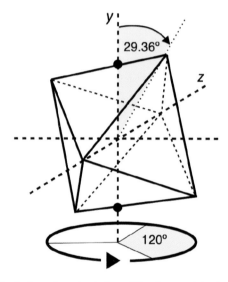

Fig. 1 the axis and rotation of the three octahedra

The accuracy of this model is within 1% of all the required angles with corresponding compensations in both modules that very nearly cancel themselves out completely. You will need 25 sheets of A4, eight each in three colours for the model itself, and one more to make the pre-folding template for the Unit B module. Cut squares from all of the sheets, giving 26 squares. Then cut only four squares of each colour (total of twelve) in half, giving 24 half-square sheets for the Unit B module as well as the 12 square sheets for the Unit A module. Cut one of the last two squares in half and use one of the resulting rectangles to make the pre-folding template. You might want to use the extra square and rectangle to 'test drive' the two modules before starting. The finished model is about 23 cm in height.

Folding the Unit A Module

1

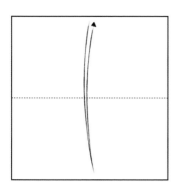

Crease the horizontal centre line.

2

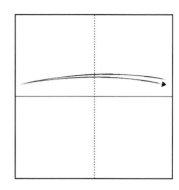

Crease the vertical centre line.

3

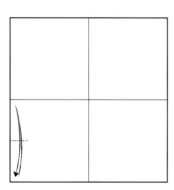

Make a short horizontal reference crease a quarter of the way up the left edge.

4

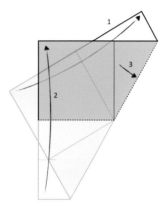

Keeping the centre fixed, fold the bottom-right corner so that the vertical centre line meets the reference crease made in step 3.

5

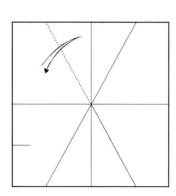

Make an inside reverse fold by bringing the bottom half upwards and folding over the vertical centre line. Then press out the resulting fold.

6

Unfold and extend the crease made in step 5 to the top edge of the module.

7

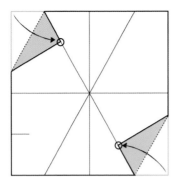

Fold the extreme top-left and bottom-right corners so that they align with the diagonal crease made in steps 5 and 6.

8

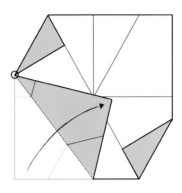

From here on, the reference crease has been removed. Fold along the line connecting the left end of the crease made in step 7 and the midpoint of the lower edge.

9

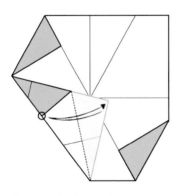

Fold the tip of the flap made in step 8 back onto itself and unfold.

10

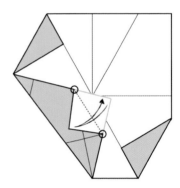

Crease from the top end of the crease made in step 9 to the end of the diagonal crease made in step 4.

11

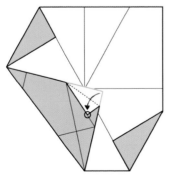

Fold the top of the flap made in step 10 back onto itself.

12

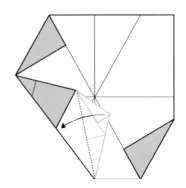

Without unfolding step 11, fold along the crease made in step 9.

13

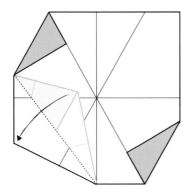

Without unfolding step 12, unfold the fold made in step 8.

14

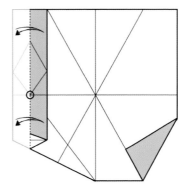

Unfold the top left corner. Using the intersection of the crease made in step 8 and the horizontal centre line as a guide, make a vertical crease.

15

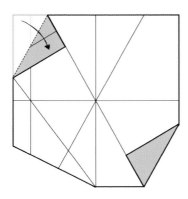

Refold the top left corner.

16

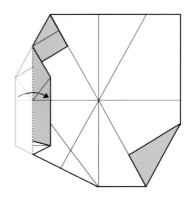

Fold the vertical crease made in step 14.

17

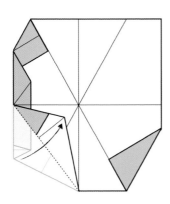

Refold the crease made in step 8.

18

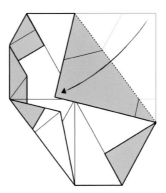

Fold along the line connecting the end of the right crease made in step 7 and the midpoint of the upper edge.

19

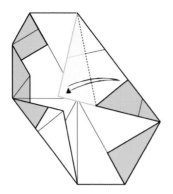

Fold the tip of the flap made in step 18 back onto itself and unfold.

20

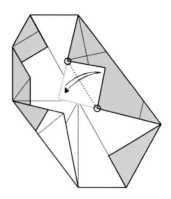

Crease from the bottom end of the crease made in step 19 to the top end of the diagonal crease made in step 4.

21

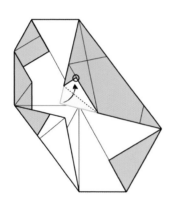

Fold the flap made in step 20 back onto itself.

22

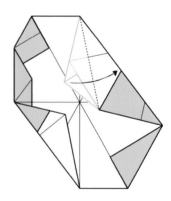

Without unfolding step 21, fold along the crease made in step 19.

23

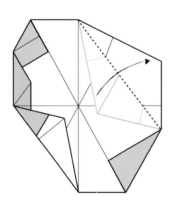

Without unfolding step 22, unfold the crease made in step 18.

24

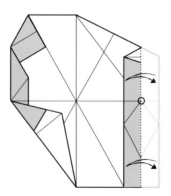

Unfold the bottom-right corner. Using the intersection of the crease made in step 18 and the horizontal centre line as a guide, make a vertical crease.

25

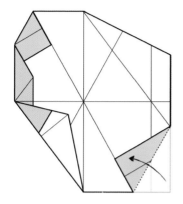

Refold the top left corner.

26

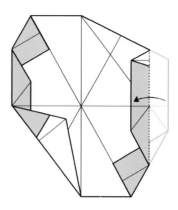

Fold the vertical crease made in step 24.

27

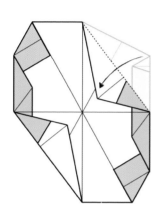

Refold the crease made in step 18.

28

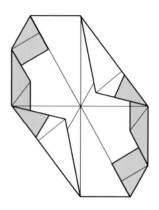

The module should now resemble the above diagram.

29

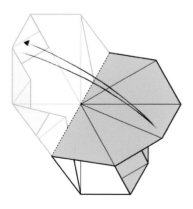

Crease the entire module along the diagonal fold made in step 4, through all layers.

30

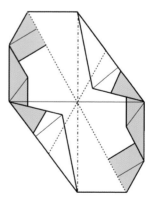

Make the following preliminary reinforcing creases: a mountain fold along the vertical centre line and creases along the two long diagonal.

31

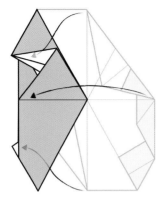

Bring the right side of the module over to the left using the creases made in step 30 to reach the above diagram.

32

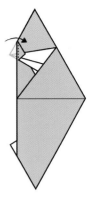

Fold the extruding upper tip back so that it clears the edge of the module.

33

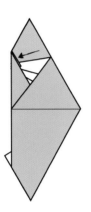

Tuck the folded tip inside the left edge.

34

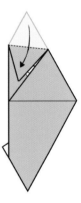

Fold the top tip of the module downwards.

35

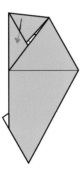

Tuck that tip also inside the left edge.

36

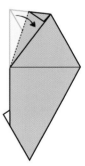

Fold the top-left flap in half. There are a good few layers to fold through so be tough with it.

37

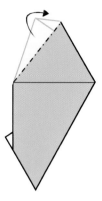

Make a mountain fold to put the entire top-left tab behind the module.

38

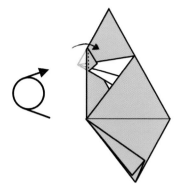

Flip the module over vertically and repeat steps 32-37 for the other side. Fold the extruding upper tip back so that it clears the edge of the module.

39

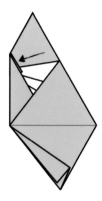

Tuck the folded tip inside the left edge.

40

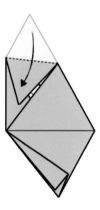

Fold the top tip of the module downwards.

41

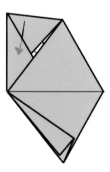

Tuck that tip also inside the left edge.

42

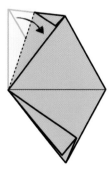

Fold the top-left flap in half. Remember to be tough with it.

43

Make a mountain fold to put the entire top-left tab behind the module.

44

Pull the module gently open and smarten up the edges. It should now resemble the above diagram.

The Pre-folding Template for the Unit B Module

The 24 Unit B modules will be folded from the half-square sheets. You should also have an extra half-square sheet, which will become a pre-folding template allowing you to make creases at 7/32 from the edges of the Unit B module. Follow the instructions below to make the pre-folding template, then pre-fold all the half-square sheets before folding the actual modules.

1

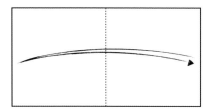

Crease the extra piece along the vertical centre-line.

2

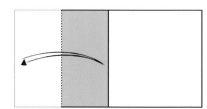

Make a vertical crease by folding the left edge of the paper to the crease made in step 1.

3

Make another vertical crease by folding the left edge of the paper to the crease made in step 2.

4

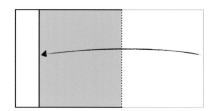

Bring the right edge over to meet the crease made in step 3 and fold the paper flat.

5

Turn the pre-folding template over so that the long side is on top.

6

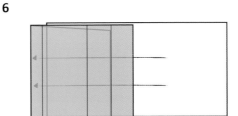

For each of the 24 Unit B modules push the paper snuggly into the template.

7

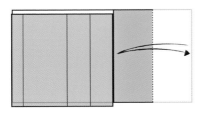

Fold the extruding edge of the module over to meet the right edge of the template and press it flat, then unfold.

8

Rotate the module 180 degrees and repeat steps 6 and 7 for the other end of the module.

Folding the Unit B Module

1

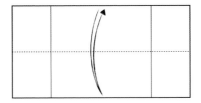

Crease the module along the horizontal centre line.

2

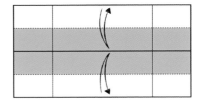

Crease the upper and lower halves in half again.

3

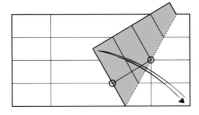

Using the right intersection of the centre line as a reference point, bring the right edge over so that the vertical crease meets the lower quarter line, then crease.

4

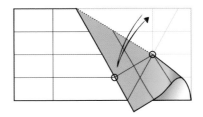

Again, using the right intersection of the centre line, fold down the top-right corner, so that the top end of the vertical crease meets the lower quarter line.

5

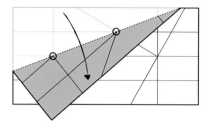

Fold the line connecting the intersection of the crease made in step 4 with the upper quarter crease and the left intersection of the centre line. Extend it to the edges.

6

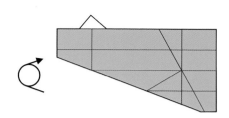

Flip the module over vertically.

7

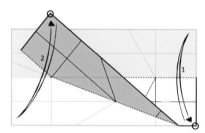

Make an inside reverse fold by bringing the upper and lower right corners together (1), and then the upper and lower left corners together (2). The fold should fall on the crease made in step 5.

8

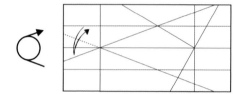

Unfold and flip the model over vertically. Extend the crease made in step 7 to the edge of the module.

9

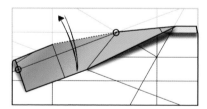

Fixing the intersection of the creases made in steps 4 and 5, bring the top-left corner down so that the top quarter line falls on the crease made in step 5. Fold only the portion between the intersection and the left vertical crease.

10

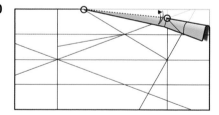

Make a crease from the intersection of the crease made in step 5 and the right vertical crease to the top end of the crease made in step 4.

11

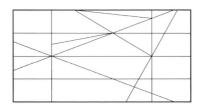

Your module should now resemble the above diagram.

12

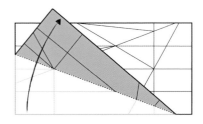

Fold along the crease made in step 7 and 8.

13

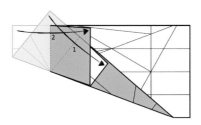

Make an inside reverse fold along the left vertical crease. Fold down the extruding tip along the left vertical crease (1), then fold the top-left corner along the left vertical crease (2).

14

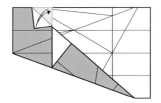

Crease the top right corner of the folded left edge down so that it still covers the edge of the flap made in step 12.

15

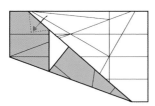

Reverse the flap made in step 14 so that to embraces all the folds made in step 13.

16

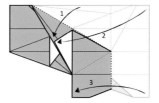

12. Make an inside reverse fold along the right vertical crease. Bring the top edge of the module down only along the existing fold (1). Bring the top-right corner across to the middle (2) then fold along the right vertical crease (3).

17

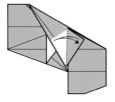

Fold the internal flap made in step 16 over to the right. Do not make this fold exactly vertical but rather allow its bottom end to swing a little to the left.

18

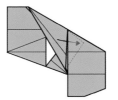

Reverse the direction of the flap made in step 17 so that it goes underneath the left vertical edge of the fold made in step 16.

19

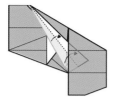

Fold along the crease made in step 10 to make a supporting diagonal 'rib'.

20

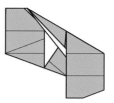

Your module should now resemble the above diagram.

21

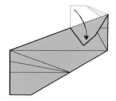

Flip the model over Vertically.

22

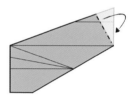

Fold the upper extruding tab back onto the module.

23

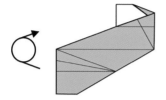

Reverse the fold made in step 22, tucking the flap in so that it embraces all the internal folds.

24

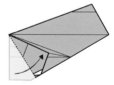

Make a mountain fold on the upper-right tip, pointing it backwards.

25

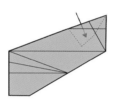

Fold and unfold the bottom-left corner along the diagonal extending down from the centre line.

26

Fold the bottom left tip across so that the lower-left tip falls exactly on the second of the three creases.

27

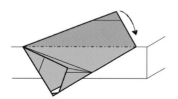

Mountain fold along the horizontal centre line. You might want to use a sharp table top or folding block to help with this fold.

28

Gently pull everything into shape so that the finished Unit B module resembles the above diagram.

Assembling the Model

There is a new joint to master in order to put this model together. It is between the long tab side of the Unit B module and the Unit A module. To make the joint you will need to partially unfold the inside of the Unit A module. Begin by pulling out one of the inside flaps made in either step 17 or 27 – they are identical so it does not matter which one (Fig. 2).

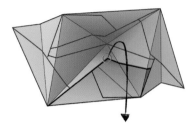

Fig. 2 unfolding step 17

Continue unfolding this flap until you can see the small flap made in step 11 (or 21). You should position the module as shown in Figure 3.

Fig. 3 revealing the small tab made in step 11

The flap made in step 11 (or 21) will be inserted under the supporting rib on the inside of the Unit B modules (Fig. 4).

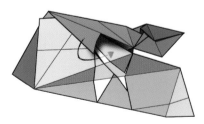

Fig. 4 insert the small tab made in step 11 under the supporting rib

Align the sharp end of the Unit B module with the corner of the Unit A module and refold the inside flap so that the modules look like figure 5.

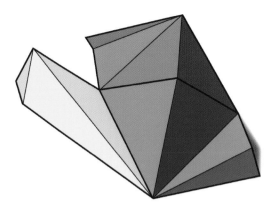

Fig. 5 the completed joint between the two units

Now add a Unit B module of the third colour to the other side in the same way. This will give you a three-coloured submodule. Make another 11 submodules, each of which include all three colours (Fig. 6).

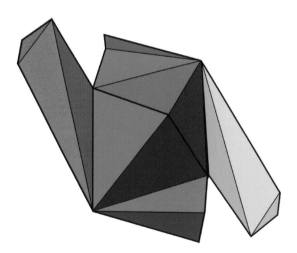

Fig. 6 one completed submodule

When you have all 12 submodules, you can start putting them together by joining the apex of four similarly-coloured Unit B modules. Make sure that the submodules have pairs of similarly-coloured Unit A modules diagonally opposite each other.

Continue to build the model, adding submodules that adhere to the colour scheme shown in the icon at the beginning of this chapter.

The Compound of Three Tetrahedra

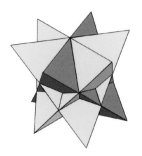

 Advanced

As far as I can tell, this is the only origami design available for this compound. One can only assume that the compound's apparently simple form and rather unbalanced proportions, unlike the many impressive origami stellated polyhedrons that abound, have dissuaded origami designers from attempting to create a model of it.

Geometrically, the form is made of three tetrahedra with a common axis. As was mentioned in chapter 2, the tetrahedron can be seen as being inscribed in a cube (Fig 1a). If that cube is rotated by 60 degrees about the central axis, the inscribed tetrahedron will now be at the position of the second tetrahedron (Fig 1.b), and a further rotation of 60 degrees will give the position of the third (Fig. 1c).

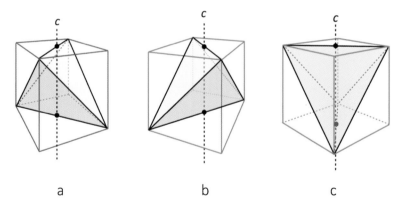

Fig. 1 the rotations of the three tetrahedra

The four modules that go together to form the model are, in reality, two pairs of mirror-image twins labeled Units A Top, A Bottom, B Left and B Right. There are a few more steps than usual for each module, so the time taken to build it might be slightly longer, but it is worth the effort. The finished model is quite sturdy and, with 21 cm × 21 cm square sheets, it stands at exactly the same height as the Compound of Three Cubes, i.e. 18.5 cm.

You will need 24 square sheets of paper. I suggest using the largest square you can get from a sheet of A4 (21 cm × 21 cm). Use three bold, modern, contrasting colours, eight squares in each. I found that stiff 80 g/m² works fine at this size.

The following diagrams show how to fold all four units. Try to be crisp and precise when folding and take care to fold only where shown. Some of the folds are of an exact length and often between reference points. Follow the diagrams carefully one by one and check each step as you go.

Folding the Unit A Module (Top and Bottom)

1

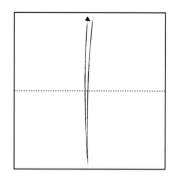

Make a crease along the horizontal centre line.

2

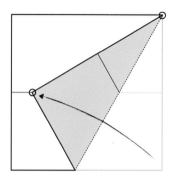

Bring the bottom-right corner up to meet the centre line and fold from the top-right corner to the bottom edge.

3

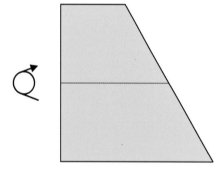

Flip the module over vertically.

4

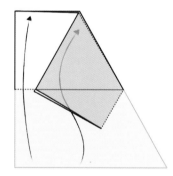

Bring the bottom-left corner up to meet the top-left corner, allowing the lower-right corner to come up to the horizontal centre line and fold.

5

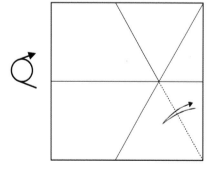

Flip the module over into its original position, then extend the crease made in step 4 to the bottom-right corner.

6

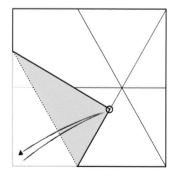

Fold the bottom-left corner up so that it lies along the fold in step 2.

7

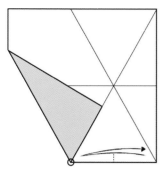

Bring the bottom-right corner over to the bottom edge of the crease made in step 6 and make a small vertical reference crease.

8

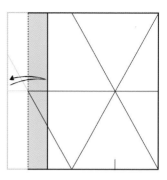

Unfold. Make a vertical crease where the horizontal centre line and the crease made in step 6 intersect.

9

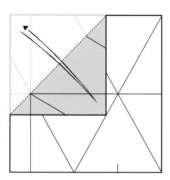

Crease the top-left corner, aligning the top of the vertical crease made in step 8 with the horizontal centre line.

10

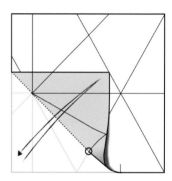

Bring up the bottom-left corner, aligning the vertical crease in step 8 and the horizontal centre line. This time, crease only as far the crease made in step 6.

11

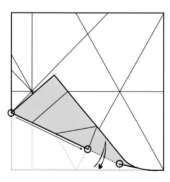

Aligning the reference point in step 10 with the left end of the crease made in step 9, crease only the portion between the reference point the crease made in step 8.

12

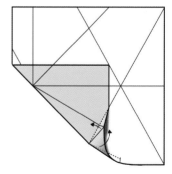

Refold step 10 and then fold along the crease made in step 2, taking in the both layers. This makes a crease on the top layer aligned with that from step 2.

13

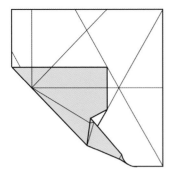

Refold steps 12 and 11. The resulting flap should resemble that in the above diagram.

14

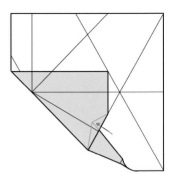

Reverse the direction of the folds made in step 13, tucking the flap inside the fold made in step 10.

15

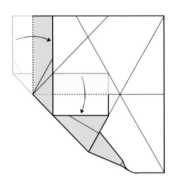

Swivel fold the top-left corner and the portion of the fold made in step 10, which lies above the horizontal centre line, clockwise.

16

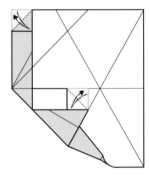

Make diagonal creases on both of the tips folded in step 15. Make these carefully since they will interlock in the finished model.

17

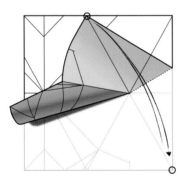

Unfold everything. Bring the bottom-right corner up to meet the other end of the crease from steps 4 and 5. Crease only the portion to the right of the intersection with the centre line.

18

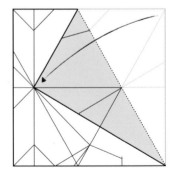

Fold along the crease made in steps 4 and 5.

19

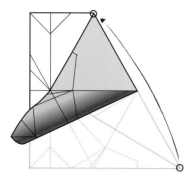

Once again, bring the bottom-right corner up to meet the other end of the crease from steps 4 and 5.

20

Keeping step 19 in place, peel back the tip of the upper layer until the crease is revealed, then fold the right half of the tip to align with the crease.

21

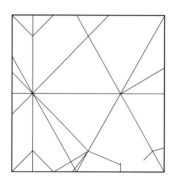

Unfold everything and compare your crease pattern with the above diagram.

22

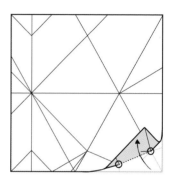

Fold between the two reference points shown above. Be careful not to fold beyond the reference points.

23

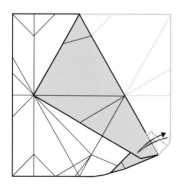

Refold the crease made in steps 4 and 5 taking in all layers.

24

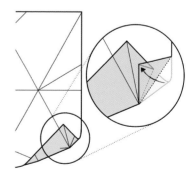

Roll over the extreme bottom-right corner and push it flat. It will find the natural position due to the single fold.

25

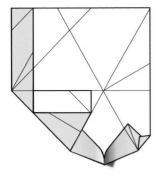

Refold the left part of the module, being
careful not to squash the tip at the
bottom edge. This tip will now determine
the final identity of the module.

Important

In the following step, the module will be defined as either Unit A Top or Unit A Bottom. The extruding tip at the bottom of the module will be folded either to the left or to the right. In step 26a the tip is folded to the left, defining the module as Unit A Top, while in step 26b the tip is folded to the right, defining the module as Unit A Bottom.

26a

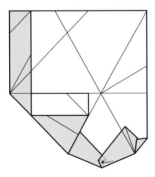

Fold the tip to the left so that the internal fold is parallel with the crease made in step 2. You may need to use a letter opener to guide the internal fold.

26b

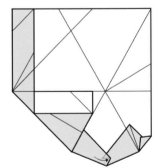

Fold the tip to the right so that the internal fold is parallel with the crease made in steps 4 and 5. You may need to use a letter opener to guide the internal fold.

From here on, the folding sequences for both the Unit A Top and the Unit A Bottom modules are exactly the same. The diagrams show the Unit A Bottom module, but when you fold the Unit A Top modules, ignore the folded tip, which will be pointing in the other direction.

You should make all six Unit A Top modules together (two each in the three colours), then put them aside and make the six Unit A Bottom modules.

27

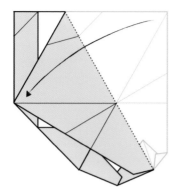

Fold along the crease made in steps 4 and 5.

28

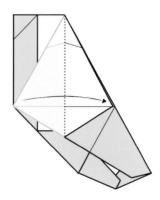

Bring the tip of the fold made in step 27 over to meet the right edge and fold vertically. Make sure it does not go over the right edge.

29

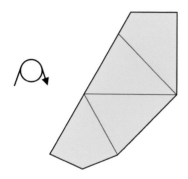

Flip the module over horizontally.

30

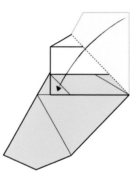

Fold over the top 'quarter', folding through all the layers. Pull the top end of the flap tightly before making the final crease.

31

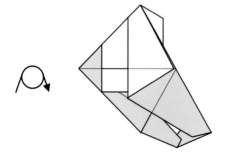

Flip the module back over horizontally.

32

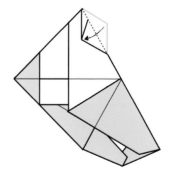

Fold the extruding flap at the top-right corner over onto the module.

33

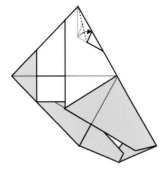

Fold half of the flap in step 32 back onto itself.

34

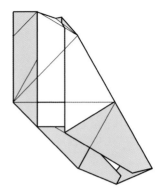

Gently unfold steps 30–33.

35

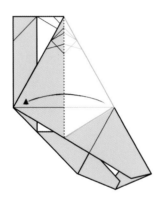

Carefully unfold the flap made in step 28.

36

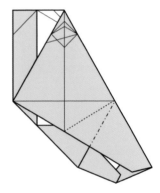

Make preparatory creases as shown above, one mountain fold and one valley fold. The remaining six diagrams use three dimensional representations.

37

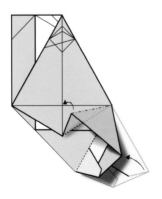

Begin 'rolling' the bottom-right corner into the middle of the module.

38

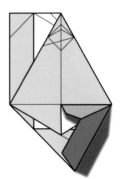

When the roll is complete, the module should look like the above diagram.

39

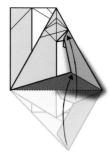

Continue the roll, but now along the horizontal centre line, until the right edge doubles up on itself.

40

Fold the flap from step 28 back inside the module. Push it carefully into place with a crochet hook or a pencil. Be careful not to pierce the module.

41

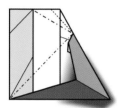

Refold steps 30–33 using the guides shown above.

42

The final Unit A module should resemble the above diagram.

The two Unit B modules are exact mirror images of each other. If you use paper that is the same colour on both sides such as coloured photocopy paper, you may make the crease pattern (steps 1–21) for all twelve modules from the following 24 diagrams for the Unit B Left Module crease pattern by flipping half of them over to fold the Unit B right module. If, on the other hand, your paper is coloured on one side only, you should follow the complete diagrams, including the crease pattern, for the Unit B Right module.

The Unit B Left Module Crease Pattern

1

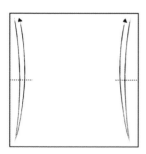

Make two horizontal reference creases at the mid points of the vertical edges. These should be about one quarter of the width of the paper.

2

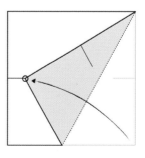

Bring the bottom-right corner up to meet the left reference crease and fold from the top-right corner to the bottom edge.

3

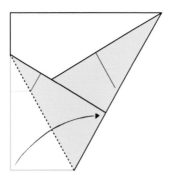

Bring the bottom-left corner up to meet the fold made in step 2 and fold. Make the lower apex as tight as possible.

4

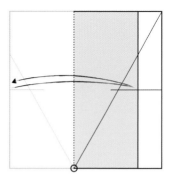

Unfold everything. Make a vertical crease, reaching from the lower apex in step 3 to the top edge.

5

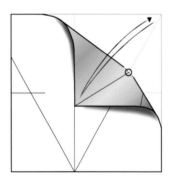

Bring the top-right corner into the middle to meet the vertical fold and crease a reference point at the intersection with the crease made in step 2.

6

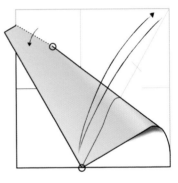

Bring the top-right corner down to the bottom of the vertical crease and fold down only the left quarter.

7

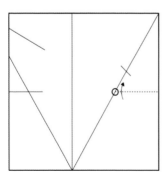

If the right horizontal crease does not already cross the long diagonal, extend it so that it does.

8

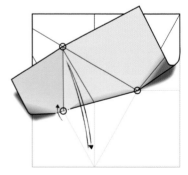

Bring the bottom end of the crease made in step 3 up to meet the crease made in step 6 and crease a reference point where the line is bisected.

9

Fold only the part to the left of the vertical crease, along the line joining the references made in steps 5 and 8.

10

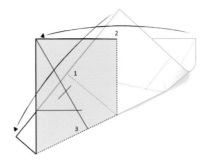

'Mirror' the crease made in step 9 in the vertical crease made in step 4, then extend it to the right edge.

11

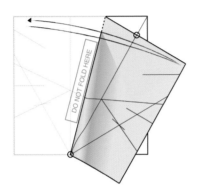

Make a reference crease by bringing the vertical crease to the long diagonal and make a small reference crease at the top edge.

12

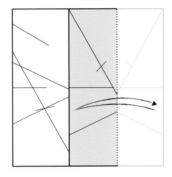

Make a vertical crease, reaching from the reference crease made in step 11 to the bottom edge.

13

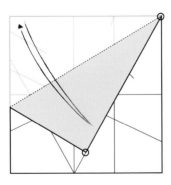

Bring the top edge down to the long diagonal made in step 2 and crease.

14

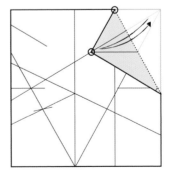

Bring the top-right corner over, aligning the tip with the crease made in step 13, while keeping the reference crease made in step 11 fixed.

15

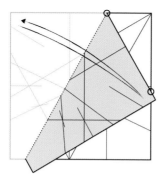

Bring the portion of the top edge to the left of the reference point from step 11 onto the crease made in step 14 and crease to the bottom edge.

16

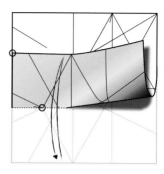

Make a horizontal crease at the height of the intersection of the creases made in steps 3 and 8, extending it to the left edge. Mark the point where it crosses the crease made in step 15.

17

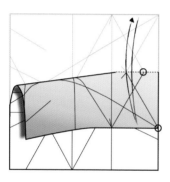

Make a horizontal crease at the height of the intersection of the creases made in steps 2 and 5, extending it to the right edge. Mark the point where it crosses the crease made in step 14.

18

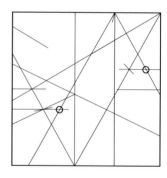

The two reference points marked in steps 16 and 17 are shown in the above diagram. Make sure you understand exactly where they are before proceeding.

19

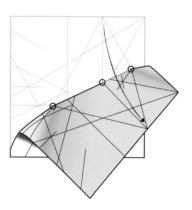

Crease only part of the line connecting these points from the one made in step 18 to the vertical crease made in step 12. Be careful not to crease any further.

20

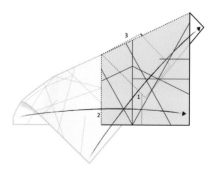

'Mirror' the crease made in step 19 in the vertical crease made in step 12.

21

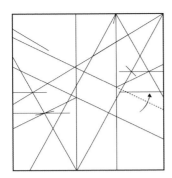

Unfold everything and extend the crease made in step 20 to the right edge. You should observe an irregular hexagon in the middle of the crease pattern.

22

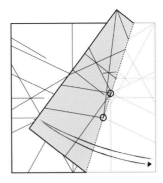

Crease along the line that connects the upper and lower-right corners of the hexagon and extend it to the top and bottom edges.

23

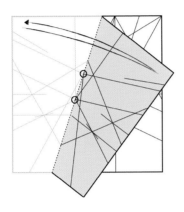

Crease along the line that connects the upper and lower-left corners of the hexagon and extend it to the top and bottom edges.

24

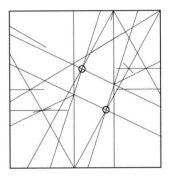

You will now see a rhombus in the middle of the crease pattern. We will crease its long diagonal.

25

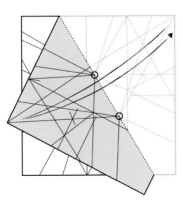

Crease the long diagonal of the rhombus and extend the crease to the upper and lower edges.

26

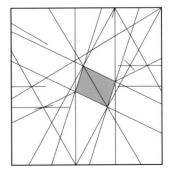

Your module should now resemble the above diagram. The shaded area is the only visible part of the module in the finished model.

The crease pattern for the Unit B Right module is a mirror image of the Unit B Left module. If you are using paper that is the same colour on both sides, then you can make all the Unit B modules from this crease pattern, turning half of them over for the Unit B Right modules and then beginning from the folding sequence on page 116.

If you are folding paper with colour on one side only, you will need to follow both sets of instructions in full, the second of which is essentially a mirror image of the first. In either case, you will make six Unit B Left modules, two in each colour and six Unit B Right modules, also two in each colour.

The Unit B Left Module Folding Sequence

1

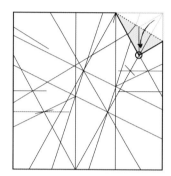

Lay the paper flat with the crease pattern as shown in the above diagram. Fold the top-right corner to fall along the 60 degree line.

2

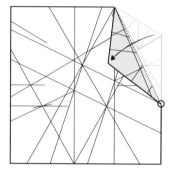

Fold along the line connecting the right end of the upper of the parallel lines and the top of the flap in step 1.

3

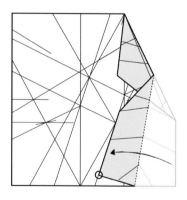

Bring the lower-right edge so that it falls along the crease extending down from the right side of the rhombus.

4

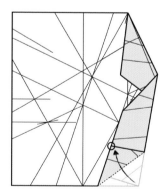

Fold the lower-right corner over so that it falls along the crease extending down from the right side of the rhombus.

5

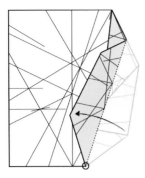

Fold along the crease extending down from the right side of the rhombus, taking in all layers.

6

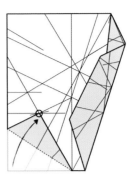

Fold the bottom-left corner up to fall along the 60 degree line.

7

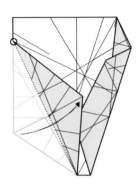

Fold along the line connecting the left end of the upper of the parallel lines and the bottom of the flap made in step 6.

8

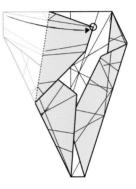

Fold the upper-left edge over so that it falls along the crease extending up from the left side of the rhombus.

9

Fold the upper-left corner over so that it falls along the crease extending up from the left side of the rhombus.

10

Fold along the crease extending up from the left side of the rhombus, taking in all layers.

11

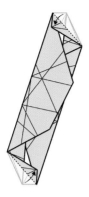

Fold the upper and lower tips of the module.

12

Tuck the tips folded in step 11 inside the module.

13

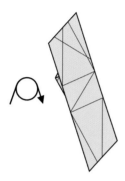

Flip the module over horizontally.

14

Crease the upper edge of the rhombus through all the layers. You can use a letter opener to hold the rhombus flat, while folding the upper part.

15

Crease the lower edge of the rhombus through all the layers. You can use a letter opener to hold the rhombus flat, while folding the lower part.

16

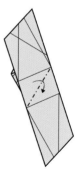

Make a mountain fold through all layers along the existing crease on the long diagonal of the rhombus.

17

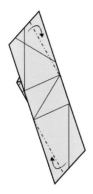

Mountain fold the upper-right and lower-left corners along the existing creases. You may want to use a hard table edge or folding block to fold these.

18

When the module is finished, it should resemble the above diagram.

The Unit B Right Module Crease Pattern

If your paper is coloured on one side only, you will need to make all the Unit B Right modules from scratch, including the crease pattern. The following diagrams are mirror images of the diagrams used in the Unit B Left module except for the colour (to help avoid confusion) and the occasional text used on the diagrams themselves. The captions for each diagram have been adjusted so that they agree with the mirror image process (i.e. left becomes right and vice versa etc). As always, take your time to compare your modules with the diagrams as you proceed. It maybe useful to mark the page where the section you're working on starts with a bookmark or a sticky memo.

1

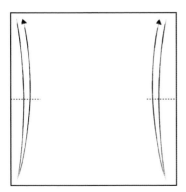

Make two horizontal reference creases at the mid points of the vertical edges. These should be about one quarter of the width of the paper.

2

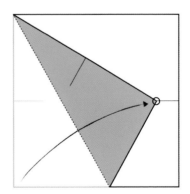

Bring the bottom-left corner up to meet the right reference crease and fold from the top-left corner to the bottom edge.

3

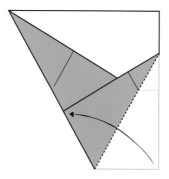

Bring the bottom-right corner up to meet the fold made in step 2 and fold. Make the lower apex as tight as possible.

4

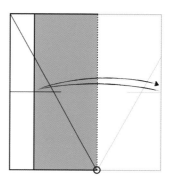

Unfold everything. Make a vertical crease, reaching from the lower apex in step 3 to the top edge.

5

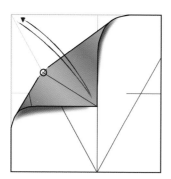

Bring the top-left corner into the middle to meet the vertical fold and crease a reference point at the intersection with the crease made in step 2.

6

Bring the top-left corner down to the bottom of the vertical crease and fold down only the right quarter.

7

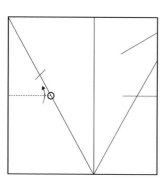

If the left horizontal crease does not already cross the long diagonal, extend it so that it does.

8

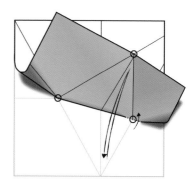

Bring the bottom end of the crease made in step 3 up to meet the crease made in step 6 and crease a reference point where the line is bisected.

9

Fold only the part to the right of the vertical crease, along the line joining the references made in steps 5 and 8.

10

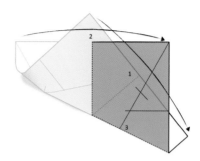

Mirror the crease made in step 9 in the vertical crease made in step 4, then extend it to the left edge.

11

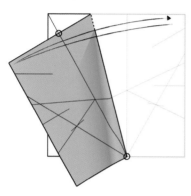

Make a reference crease by bringing the vertical crease to the long diagonal and make a small reference crease at the top edge.

12

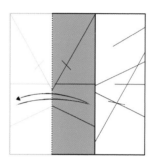

Make a vertical crease, reaching from the reference crease made in step 11 to the bottom edge.

13

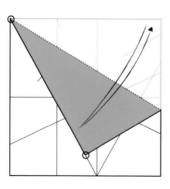

Bring the top edge down to the long diagonal made in step 2 and crease.

14

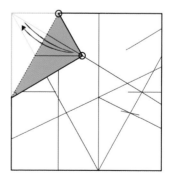

Bring the top-left corner over, aligning the tip with the crease made in step 13, while keeping the reference crease made in step 11 fixed.

15

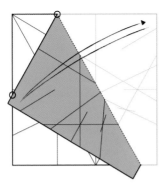

Bring the portion of the top edge to the right of the reference point from step 11 onto the crease made in step 14 and crease to the bottom edge.

16

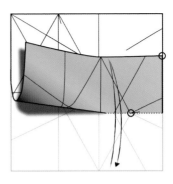

Make a horizontal crease at the height of the intersection of the creases made in steps 3 and 8, extending it to the right edge. Mark the point where it crosses the crease made in step 15.

17

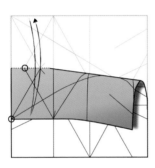

Make a horizontal crease at the height of the intersection of the creases made in steps 2 and 5, extending it to the left edge. Mark the point where it crosses the crease made in step 14.

18

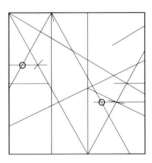

The two reference points marked in steps 16 and 17 are shown in the above diagram. Make sure you understand exactly where they are before proceeding.

19

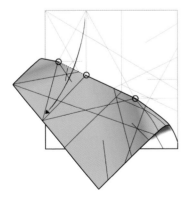

Crease only part of the line connecting these points from the one made in step 18 to the vertical crease made in step 12. Be careful not to crease any further.

20

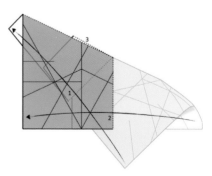

'Mirror' the crease made in step 19 in the vertical crease made in step 12.

21

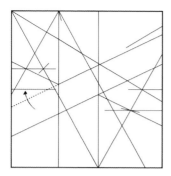

Unfold everything and extend the crease made in step 20 to the left edge. You should observe an irregular hexagon in the middle of the crease pattern.

22

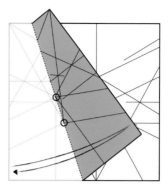

Crease along the line that connects the upper and lower-left corners of the hexagon and extend it to the top and bottom edges.

23

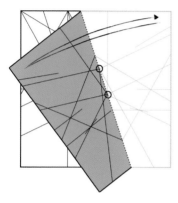

Crease along the line that connects the upper and lower-right corners of the hexagon and extend it to the top and bottom edges.

24

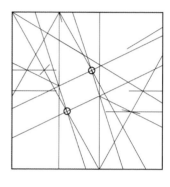

You will now have a rhombus in the middle of the crease pattern. We will crease its long diagonal.

25

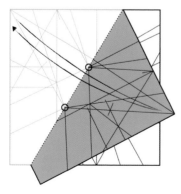

Crease the long diagonal of the rhombus and extend the crease to the upper and lower edges.

26

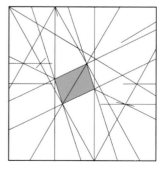

Your module should now resemble the above diagram. The shaded area is the only visible part of the module in the finished model.

> **Top Tip: Always read the entire chapter thoroughly before beginning a model. Never make a fold without first understanding how it should end up. Always look ahead at the next couple of diagrams to see how things should end up and see how the current fold is incorporated. Make an effort to understand how that fold is essential to the finished design. In short, try to get inside the designer's mind!**

The Unit B Right Module Folding Sequence

1

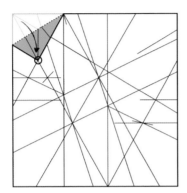

Lay the paper flat with the crease pattern as shown in the above diagram. Fold the top-left corner to fall along the 60 degree line.

2

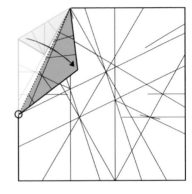

Fold along the line connecting the left end of the upper of the parallel lines and the top of the flap in step 1.

3

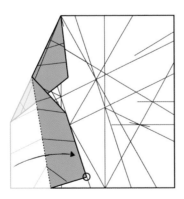

Bring the lower-left edge so that it falls along the crease extending down from the left side of the rhombus.

4

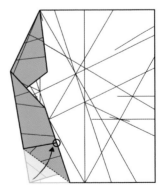

Fold the lower-left corner over so that it falls along the crease extending down from the left side of the rhombus.

5

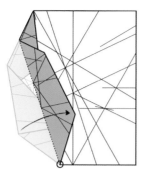

Fold along the crease extending down from the left side of the rhombus, taking in all layers.

6

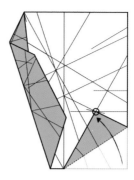

Fold the bottom-right corner up to fall along the 60 degree line.

7

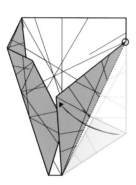

Fold along the line connecting the right end of the upper of the parallel lines and the bottom of the flap made in step 6.

8

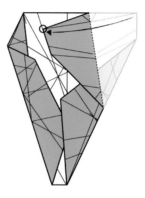

Fold the upper-right edge over so that it falls along the crease extending up from the right side of the rhombus.

9

Fold the upper-right corner over so that it falls along the crease extending up from the right side of the rhombus.

10

Fold along the crease extending up from the right side of the rhombus, taking in all layers.

11

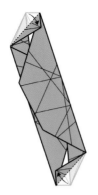

Fold the upper and lower tips of the module.

12

Tuck the tips folded in step 11 inside the module.

13

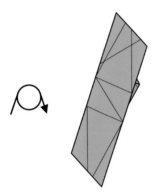

Flip the module over horizontally.

14

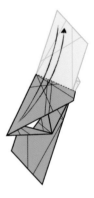

Crease the upper edge of the rhombus through all the layers. You can use a letter opener to hold the rhombus flat, while folding the upper part.

15

Crease the lower edge of the rhombus through all the layers. You can use a letter opener to hold the rhombus flat, while folding the lower part.

16

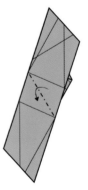

Make a mountain fold through all layers along the existing crease on the long diagonal of the rhombus.

17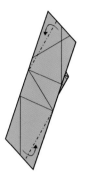

18

Mountain fold the upper-left and lower-right corners along the existing creases. You may want to use a hard table edge or a folding block to fold these.

When the module is finished, it should resemble the above diagram.

Assembling the Model

Before we start, a little explanation of the anatomy of the modules is in order. Unlike most of the other models, this one uses a complex combination of locks, slots and interweaving to hold it together. Figure 2 shows the inside of both modules, labelling the various features used in the assembly.

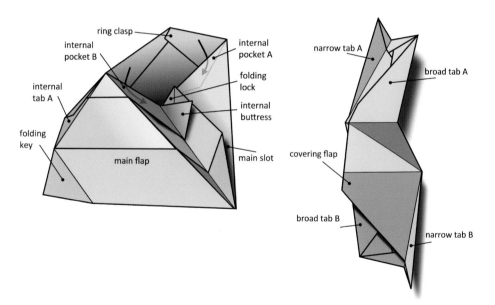

Fig. 2 the inner anatomy of the modules

The joints will be explained during the assembly, using the terminology in Figure 2. This one is complicated. Work slowly so you don't miss anything.

The Bottom Carrousel

You should begin by assembling the bottom carrousel, which is made up of the six Unit A Bottom modules. After ensuring that both the folding lock and key are unfolded, push the main tab into the main slot of the neighbouring module. As you do that, insert internal tab 'A' into the internal pocket 'A'. When these are in place press the connected sides of the two modules together and hold them in place with a paper clip. Push the internal buttress up to reveal the folding lock. Then, with a medium-sized crochet hook, pull the folding lock diagonally outwards. When the lock is folded completely, bring the internal buttress down to its original position and press the key shut. Remove the paper clip and repeat for the remaining five Unit A Bottom modules to create the carrousel below. When you have constructed the bottom carrousel, place it on the worktop with the its innards facing upwards (Fig. 3).

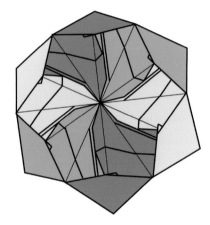

Fig. 3 the bottom carrousel

The Zig-Zag Ring and Top Carrousel

The Unit B and Unit A Top modules are added to the bottom carrousel in the following order.

Insert the narrow tab A of a Unit B Left module into the internal pocket A of the module on the anti-clockwise side of the Unit A module with the same colour. The narrow tab should slide in behind the main flap of the Unit A module with the same colour (Fig. 4a).

Insert the broad tab B of a Unit B Right module into the internal pocket B of the Unit A module that has the same colour as the B Module in step 1. The broad tab should go under the ring clasp, while the covering flap of the Unit B in step 1 goes over the broad tab B of the second Unit B module (Fig. 4b).

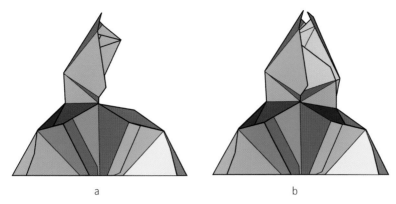

a   b

Fig. 4 inserting the Unit B modules

Now add the Unit A Top module with the remaining colour to the combination of the two inserted Unit B modules making sure the ring clasp is in place.

The next step is to add the next clockwise Unit A Top module to the first. Use exactly the same process as you did for the bottom carrousel. The only difference is that you should not remove the paper clip until the next two Unit B modules are in place. Only when you have got those in place should you remove the paper clip and work the tabs of these into place in the second Unit A Top module, making sure the ring clasp takes in the edge of the broad tab.

Work clockwise round the model, adding first the Unit A Top modules then the corresponding two B Units. When you get to the second last Unit A Top module, add the last one too. This might be tricky since it will be almost 'blind' folding, so get very familiar with the folding lock on a couple of spare modules before trying this on the model.

Only once you have the last top unit in place, should you add the final Unit B modules. I suggest putting these completely inside the model before slotting them into place.

This is the most complex model in the book, and it is not easy to avoid knocking the model about while building it. Lots of light and much patience is demanded, but the result is a stunning geometrical model which is also very robust.

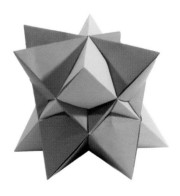

The Compound of the Icosahedron and the Dodecahedron

 Easy Intermediate

The two largest Platonic solids, the Regular Icosahedron and the Regular Dodecahedron, combine to create a beautiful form with 120 faces – sixty equilateral triangles and sixty isosceles triangles with 36 degree corners.

Like the Compound of Cube and Octahedron earlier in this book, the two solids coincide at the midpoints of all their edges, forming raised pyramids that correspond to the apexes of the original solids. This is a result of a special relationship between certain pairs of polyhedra. Every polyhedron has a corresponding 'dual' – such pairs are called **dual pairs**.

In this model, the angles of the isosceles triangle, which ideally should be 108 and 36 degrees have been approximated using the 8 × 11 right-angled triangle. This gives an angle of 36.02. Below, you can see how this triangle is created in the module at step 6, using eleven and eight sixteenths.

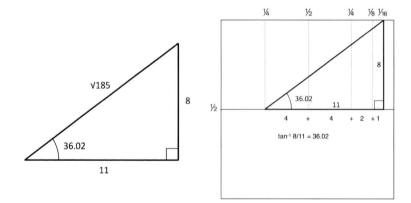

To make this model, you will need 60 square sheets of paper. Cut 18 sheets of 80 g/m² A4, (three each of six different colours) into four squares each.

Follow the instructions on the next pages carefully and, as always, check your result with the diagrams before moving on.

Folding the Module

1

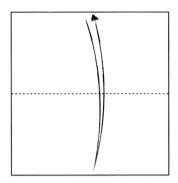

Crease the horizontal centre line.

2

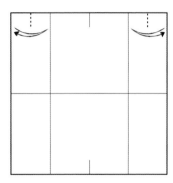

Make reference creases at the midpoints of the top and bottom edges.

3

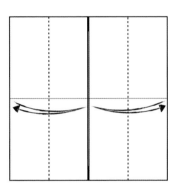

Use the reference creases to fold and unfold the left and right quarters.

4

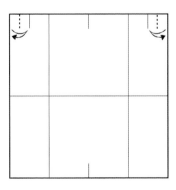

Make two more reference creases on the top edge one eighth from the left and right edges.

5

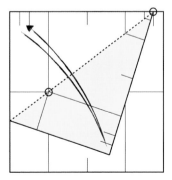

Once more, make two reference creases on the top edge, now one sixteenth from the left and right edges.

6

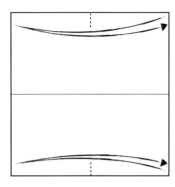

Crease from the one-sixteenth right reference crease to the left edge, passing through the intersection of the centre line and the left quarter crease.

7

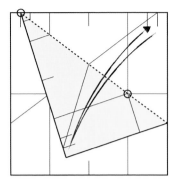

Crease from the one-sixteenth left reference crease to the right edge, passing through the intersection of the centre line and the right quarter crease.

8

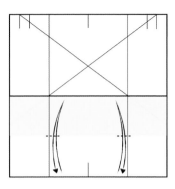

Make short reference creases on the quarter creases one quarter from the bottom edge by bringing the bottom edge up to the centre line.

9

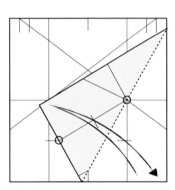

Bring the bottom-right corner up so that the end of the right quarter crease falls on the mark made in step 8, while fixing the intersection of the centre line and the right quarter crease.

10

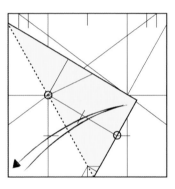

Bring the bottom-left corner up so that the end of the left quarter crease falls on the mark made in step 8, while fixing the intersection of the centre line and the left quarter crease.

11

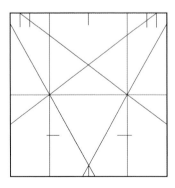

Compare your module with the above crease pattern. In the subsequent diagrams, all of the small reference creases have been omitted.

12

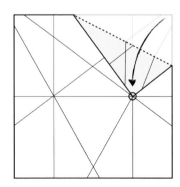

Bring the top-right corner down to the intersection of the centre line and the right quarter crease, and fold.

13

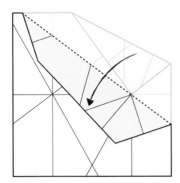

Fold along the crease made in step 7, with the fold made in step 12 inside.

14

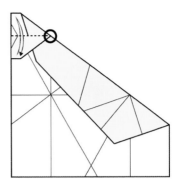

Crease the top-left corner horizontally at the point where the fold made in step 13 intersects the left vertical quarter crease.

15

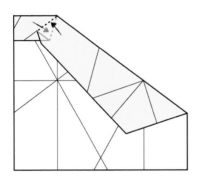

Invert step 14 by bringing the large flap out on top. Use the crease made in step 14 as the internal fold and press out the fold.

16

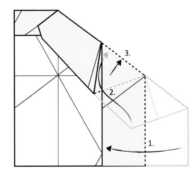

Make a reverse inside fold using the right vertical quarter crease and the diagonal made in step 7. Press out the top fold (3).

17

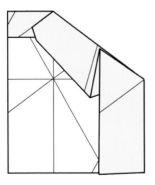

Before continuing, check that your module looks like the above diagram. Remember that the reference creases have been omitted.

18

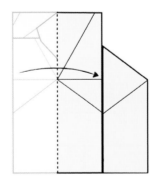

Fold the left quarter over completely, taking in all the layers.

19

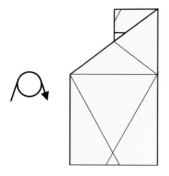

Flip the module over horizontally.

20

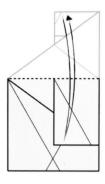

Fold and unfold the entire module along the horizontal centre line, folding through all layers.

21

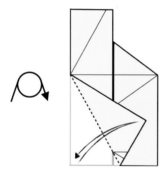

Flip the module over horizontally, and fold the bottom-left corner over along the crease made in step 10.

22

Invert the fold made in step 21 and tuck it inside the fold made in step 16.

23

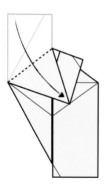

Fold the top-left part of the module down along the crease made in step 6.

24

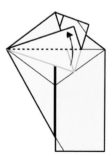

Fold up the bottom corner of the flap made in step 23 back so that the horizontal centre line is revealed.

25

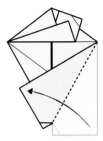

Fold the bottom-right corner over along the crease made in step 9.

26

Fold the extruding tip at the bottom of the module upwards.

27

Tuck the tip inside the module.

28

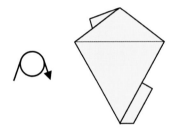

Flip the module over horizontally.

29

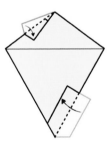

Fold the extruding flaps over on to the module.

30

Ease the module gently open so that it resembles the above diagram. Now make another 59 of them.

Assembling the Model

Once you have folded all sixty modules, you can begin assembling the model. A bit of forward planning is required, since the best approach is to construct 20 triangular sections from three modules and then to join them together like an icosahedron. This means that you must be careful to join the right coloured pieces together to make the triangular sections.

Triangular Sections

Begin by taking three modules of different colours. Push the tabs into the corresponding slots to make an equilateral triangle with a small pyramid on it (Fig. 2). Ease the three modules together and gradually tighten them.

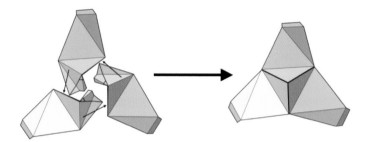

Fig. 2 constructing the equilateral triangular sections from three modules

Pentagonal Sections

Put together another four triangular sections following the colour scheme below. connect them together to form one large pentagonal section (Fig. 3).

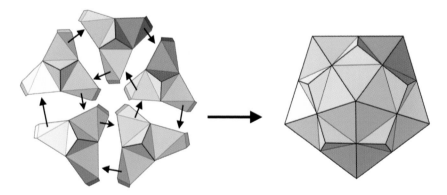

Fig. 3 five triangular sections go together to form one pentagonal section

Finishing the Model

As you make and add more triangular sections, use the following two rules to find the right colours.

1. The modules of the triangular sections always have the same colour as their neighbour.
2. The third colour is always the farthest away colour on the large pentagonal section to which they are attached (Fig 4).

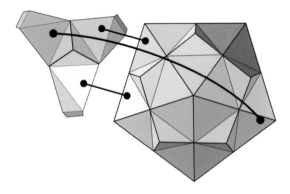

Fig. 4 two rules for determining the colour scheme of the triangular sections

The Small Triambic Icosahedron

 Easy Intermediate

The mysterious sounding word in the title, triambic, identifies a special property of the First Stellation of the Icosahedron. A **triambus** is any equilateral hexagon, other than the familiar Regular Hexagon, which maintains a triangular symmetry. Fig. 1 shows four equilateral hexagons. the first is the Regular Hexagon; hexagons b and c are both triambi, while hexagon d is an irregular equilateral hexagon with no triangular symmetry and therefor is not a triambus.

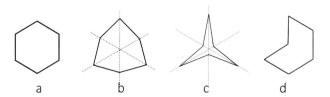

Fig. 1 Four hexagons

The First Stellation of the Icosahedron can be seen as being made up of 20 intersecting triambi. Fig. 2a shows two isolated intersecting triambi and how they fit into the scheme of the Small Triambic Icosahedron.

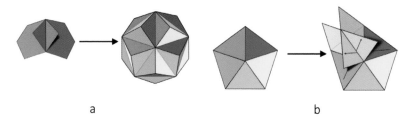

Fig. 2 Intersecting polygons (a) compared with stella1on (b)

The same form can be reached by applying the process of **stellation** to the Icosahedron. In stellation, all the faces of a polyhedron are extended to the point where they intersect each other. Fig. 2b shows this extension applied only to one edge of the blue and purple faces. If this is applied to all the faces then the result is the First Stellation of the Icosahedron.

To make this model, you will need 31 square sheets of paper. Cut 15 pieces of 80 g/m² A4 into two squares each – three each in five different colours. Cut one more square for the pre-folding template.

The pre-folding template allows you to make a crease three sixteenths for the edge of the module. The crease facilitates an excellent approximation for the exact angle of 38.3 degrees, using the $4-5-\sqrt{41}$ triangle to give 38.6 degrees with about 1% error.

Follow the instructions on this and the next pages carefully and, as always, check your result with the diagrams before moving on.

Making and Using the Pre-folding Template

1

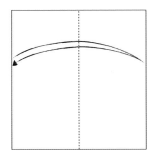

Crease the vertical centre line.

2

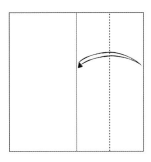

Crease the right half in half again.

3

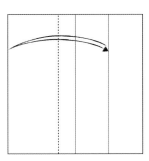

Bring the left edge to the crease made in step 2 and crease.

4

Flip the template over so that the long side is on top. Now it's ready for use.

5

Push the module snuggly into the template.

6

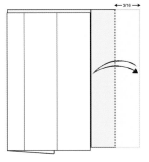

Bring the right edge of the module over to meet the right edge of the template.

7

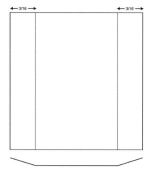

Repeat for the left edge. The pre-folded module should look like the above diagram.

Folding the Module

1

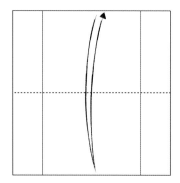

Crease the horizontal centre line.

2

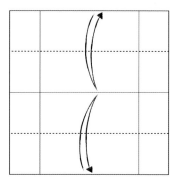

Crease the upper and lower halves, dividing the module into quarters.

3

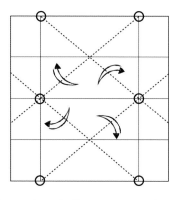

Crease along the lines connecting the reference points shown in the above diagram.

4

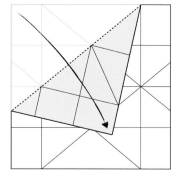

Fold down the top-left corner along the top-left crease made in step 3.

5

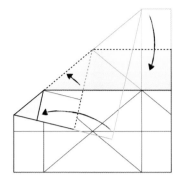

Make an inside reverse fold on the top quarter.

6

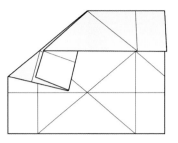

The result should look like the above diagram.

7

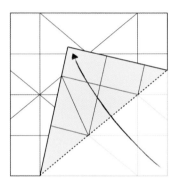

Unfold everything. Now fold the bottom-right corner along the bottom-right crease made in step 3.

8

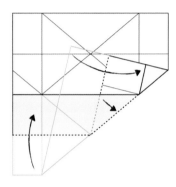

Make an inside reverse fold on the bottom quarter.

9

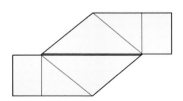

Refold steps 4 and 5, tucking the folds inside to reach the diagram above.

10

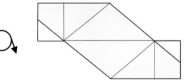

Flip the module over horizontally.

11

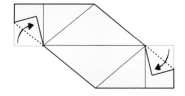

Fold over the extruding flaps.

12

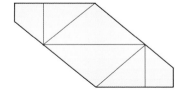

Tuck the extruding flaps inside the module, so as to match the above diagram.

13

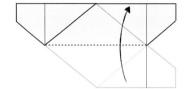

Fold the bottom half of the module upwards.

14

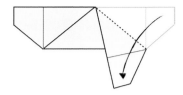

Fold the top-right corner down along the diagonal crease.

15

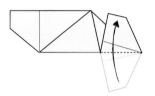

Fold the flap back up so that crease falls along the bottom edge of the module.

16

Fold the extruding tip back down to fall on top of the module.

17

 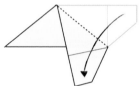

Flip the module over horizontally and fold the top-right corner down along the diagonal crease.

18

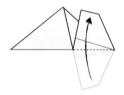

Fold the flap back up so that crease falls along the bottom edge of the module.

19

Fold the extruding tip back down to fall on top of the model.

20

Pull the module gently out so that it looks like the above diagram. Make another 29 of them.

Assembling the Model

The model is assembled by simply inserting the tabs into the respective slots in the other modules. Begin by joining three modules of different colours to form a triambus with a raised triangular pyramid.

Continue by adding two more modules of the remaining two colours to form a second triambus.

Add two more modules to create one of the pentagonal junctions, using the following rule to create the colour scheme: The modules that are diametrically and perpendicularly opposed across the pentagonal junction have the same colour (Fig. 3).

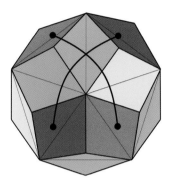

Fig. 3 Rule for creating the colour scheme

As you add more modules to the model you will find that it will be increasingly hard to get the tab into the respective slot. To avoid damaging the modules you should loosen them slightly. Then, when the new module moves in place you should tighten the neighbouring modules, too. You will also find that the modules pop up into a curved shape, allowing the tabs to slide more easily into their respective slots.

The Compound of the Small Triambic Icosahedron and the Dodecahedron

 Intermediate Advanced

The final model in this book takes us in to the realm of **irregular polyhedron compounds**, which means that at least one of the polyhedra involved is irregular. This particular model combines the Dodecahedron and the First Stellation of the Icosahedron (also known as the Small Triambic Icosahedron or STI). The five-pointed stars embossed on each of the dodecahedral faces are formed by the STI protruding beyond the limits of the Dodecahedron.

The paper used in the two modules are of different sizes. The Unit B module is three quarters of the size of the Unit A module. You will need 15 sheets of A4 for the 30 Unit A modules and another 16 sheets of A4 in a contrasting colour for the 60 Unit B modules and the two pre-folding templates. Use the stiffest 80 g/m² paper you can find. Cut 62 squares 10.5 × 10.5 cm from the 15 sheets in the colour you want the the stars to be and 30 squares 14 × 14 cm from the other 15 sheets (Fig. 1).

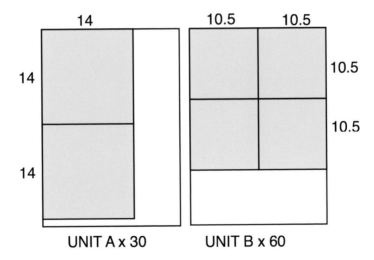

Fig. 1 the A4 cutting guide for the Unit A and Unit B modules with ratio 4:3

Making the Pre-folding Templates

Carefully follow the instructions on this and the next page to make the pre-folding templates from the two 'extra' squares the same size as those used for the Unit B module.

The 'One Seventh' Template

1

Crease along the horizontal centre line.

2

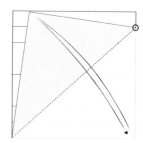

Bring the upper and lower edges to the centre line, crease and unfold.

3

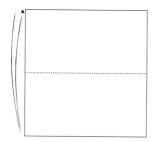

Bring the upper and lower edges to the creases made in step 2, crease and unfold.

4

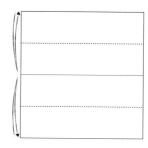

Crease from the bottom-left corner to the right end of the upmost horizontal crease.

5

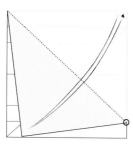

Crease from the top-left corner to the right end of the lowest horizontal crease.

6

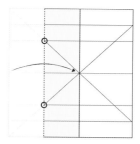

Fold over the left two-sevenths of the model, using the reference points shown in the above diagram as guides.

7

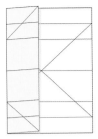

The finished template should look like the above diagram.

The 'One Third' Template

1

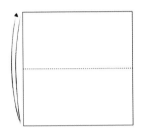

Crease along the horizontal centre line.

2

Bring the upper and lower edges to the centre line, crease and unfold.

3

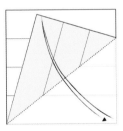

Crease from the bottom-left corner to the right end of the upper horizontal crease.

4

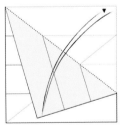

Crease from the top-left corner to the right end of the lower horizontal crease.

5

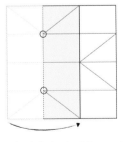

Fold over the left third of the model, using the reference points shown in the above diagram.

6

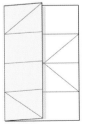

The finished template should look like this.

Put both of these templates to one side until you begin making the Unit B modules.

Now that the preliminaries have been taken care of, it's time to make the 30 Unit A modules. As with all the other models in this book, take your time and follow the instructions carefully, comparing your results with the diagrams as you go. Don't be tempted to build the model as you make the modules. This particular model is quite heavy and having it wait around for the next module to appear while supporting fifty modules is asking for disaster. Be patient and the build will go smoothly with all the modules ready at hand.

Folding the Unit A Module

1

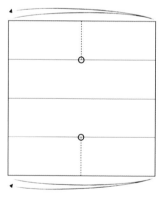

Crease along the horizontal centre line.

2

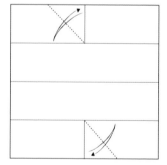

Bring the upper and lower edges to the centre line and crease.

3

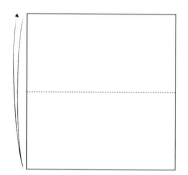

Make vertical creases between the upper and lower edges and the creases made in step 2, at the horizontal mid-point.

4

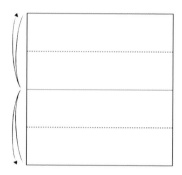

Make diagonal creases from the intersections of the creases made in steps 2 and 3 towards the upper and lower edges.

5

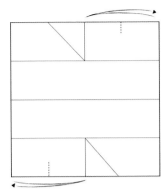

Make small reference creases on the top and bottom edge at a quarter of the length, on the other side from the diagonal crease.

6

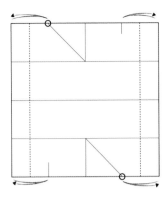

Fold both vertical edges in to meet the reference crease and the end of the diagonal crease and crease them.

7

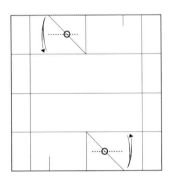

Fold over the upper and lower quarters and crease only at the point where the fold crosses the crease made in step 4.

8

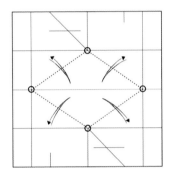

Crease the four diagonals between the nodes shown in the above diagram.

9

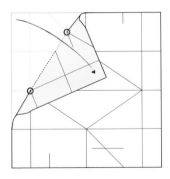

Fold along the line connecting the point marked in step 7 and the left most node in step 8. Crease only the lower marked portion of the fold.

10

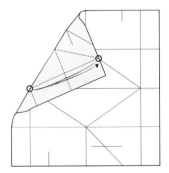

Fold over the top half of the flap made in step 9, bringing the two points indicated in the diagram together. Press flat.

11

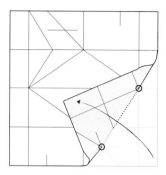

Repeat step 9 but now for the diagonally opposite corner.

12

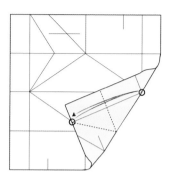

Repeat Step 10 for the diagonally opposite corner.

13

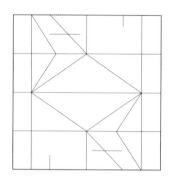

Unfold everything and check that your crease pattern looks like the diagram above.

14

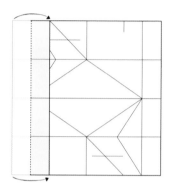

Fold over the extreme left eighth of the module.

15

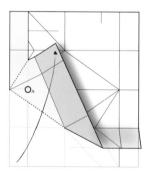

Keeping the middle of the previous fold in place, fold the bottom left corner up while pulling the eighth out to make an inside reverse fold.

16

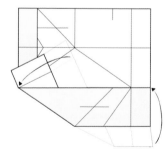

Pull the top of the flap made in step 15 down diagonally and bring the lower edge up to the horizontal centre-line.

17

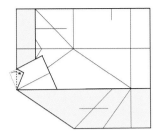

Fold the corner that crosses the left edge back over onto the module.

18

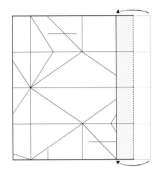

Unfold everything and fold over the extreme right eighth of the module.

19

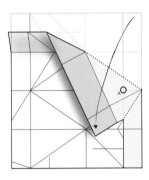

Keeping the middle of the previous fold in place, fold the top-right corner down while pulling the eighth out to make an inside reverse fold.

20

Pull up the bottom of the flap made in step 19 diagonally and bring the upper edge down to the horizontal centre-line.

21

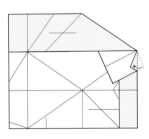

Fold the corner that crosses the right edge back over onto the module.

22

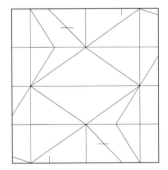

Unfold the model and check that the crease pattern looks like the diagram above.

23

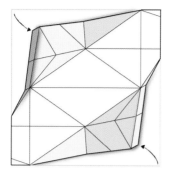

The next three diagrams should be performed as a single movement. Lift up the lower-right and upper-left corners, gently pushing the one eighth flaps over as you go.

24

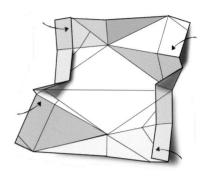

Continue to push the corners inwards and mountain fold the upper-right and lower-left sections of the module as you go.

25

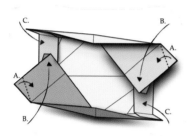

A. Fold over the triangular excesses on both sides.

B. Bring the top and bottom horizontal quarters of the model into the centre line.

C. Tuck the mountain folds under the top and bottom quarters.

26

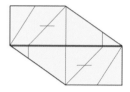

When the maneuver described in steps 23-25 are complete, the module should look like the diagram above.

27

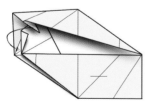

Lift up the top quarter at the left side and tuck the inner flap under all the folds on the lower half.

28

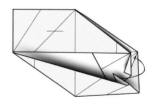

Repeat step 27 for the lower quarter.

29

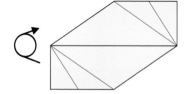

Flip the module over vertically.

30

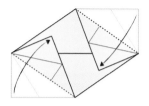

Lift the entire lower-left flap upwards along the existing crease, then repeat for the upper-right flap.

31

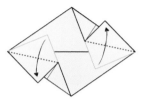

Bring the tips of the flaps back across folding along the existing creases.

32

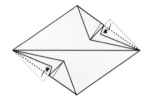

Fold the excess of the flaps back onto the module.

33

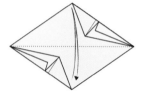

Finally, reinforce the centre line of the module.

34

Flip the module over vertically and pull the tabs out gently. Final Unit A modules should resemble the above diagram.

Once you have completed folding the 30 Unit A modules, put them to one side and bring out the two pre-folding templates for the Unit B module. Make the preliminary creases in all 60 of these modules before moving on to the folding sequence. During the folding sequence you will need to have the one third template on hand, but there you will use it to fold one eighth rather than one third. For this, you will use the half and quarter folds as a 'one eighth' template. In the meantime, let's get make the preliminary creases in 60 Unit B modules.

Pre-folding the Unit B Module

1

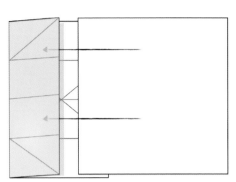

Begin by inserting the paper into the one-third pre-folding template.

2

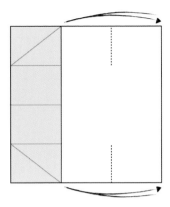

Make two short creases no longer than a quarter of the width of the module, one from the top edge and one from the bottom edge.

3

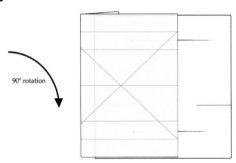

Rotate the paper 90 degrees clockwise and now insert it into the one-seventh pre-folding template.

4

Make two tiny reference creases one-seventh of the distance from both sides of the top edge.

5

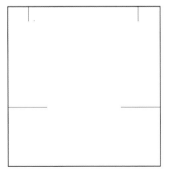

Your module should now resemble the above diagram.

Folding the Unit B Module

1

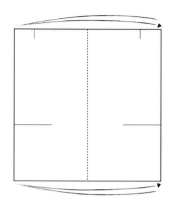

Make a crease down the vertical centre line.

2

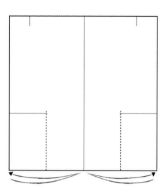

Make two vertical creases one third of the height of the paper by halving again the two vertical halves. Extend the one-third creases to intersect.

3

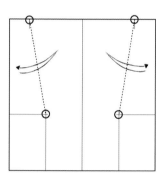

Make creases joining the intersections created in step 2 with the one-seventh reference creases made in step 4 of the pre-folding instructions.

4

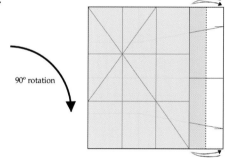

90° rotation

Rotate the model clockwise by 90 degrees. Then use the side folded in quarters on the 'one-third' template to fold over one eighth of the right edge.

5

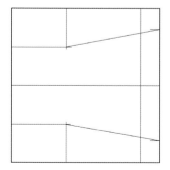

Check that your module now resembles the above diagram.

6

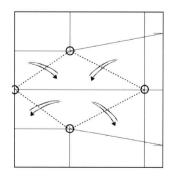

Make creases that join the four reference points shown in the above diagram to form a kite shape.

7

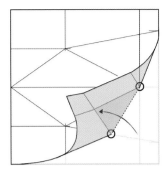

Bring the vertical crease made in step 4 onto the lower-right crease made in step 6. Press the fold from the rightmost reference point as far as the crease made in step 3.

8

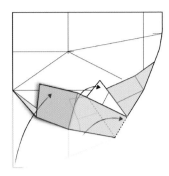

Fold the remainder of the crease made in step 3 and press flat the lower-right edge. Note that the lower-left corner will rise up.

9

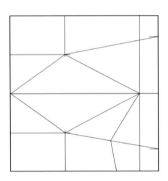

Unfold everything and check that your module now looks like the above diagram.

10

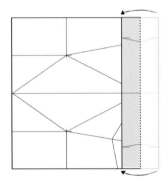

Fold over the right eighth of the module.

11

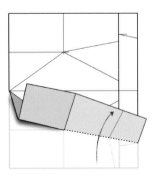

Fold up the lower edge of the module along the crease made in step 3, making sure to fold the layers inside the eighth flap.

12

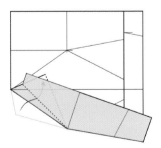

Fold the raised bottom-left corner inside, using the crease made in step 6, then flattening out the top of the resulting flap.

13

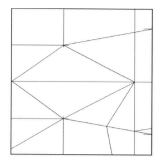

Once again, unfold the module and check that it now looks like the above diagram.

14

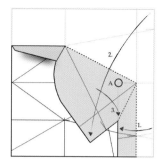

1. Refold the right eighth.
2. While keeping the mid point fixed at 'A', bring the top-right corner down.
3. Open out and flatten the upper half of the eighth flap and push the fold flat.

15

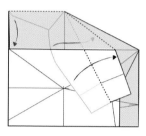

Bring the lower edge of the flap made in step 14.2 across to the right and fold the top-left corner down to the centre line.

16

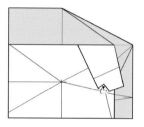

Fold the portion of the flap created in step 15 upwards so it clears the crease made in step 3.

17

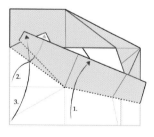

Refold step 12.

18

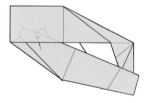

Tuck the top of the flap made in step 17 underneath the flap made in step 15.

19

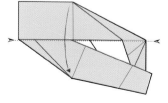

Fold the entire module along the centreline. A folding block may be useful in making this fold.

20

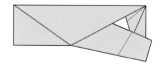

After folding the centre-line, press the module flat.

21

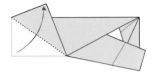

Fold back the bottom-left corner along the existing fold.

22

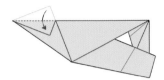

Fold the excess of the flap made in step 25 back down onto the module.

23

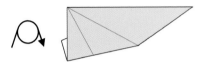

Flip the module over horizontally.

24

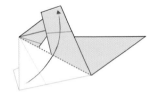

Fold the whole of the bottom-left flap up along the existing fold.

25

Bring the top of the flap made in step 28 back down along the existing fold.

26

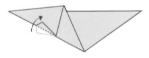

Fold the excess of the flap back up onto the model.

27

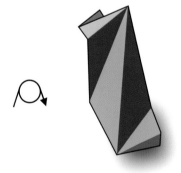

Gently pull the unit open and check that it looks like the above diagram.

When you have completed all 60 of the Unit B modules, take a well earned rest. Gather your nerves and get ready to move on to the assembly stage. You will need a lot of patience and a pad of sticky memos like Post-It or StickNote. I cut of nearly all of the non-sticky part of these so it doesn't get in the way but leave enough so you can pick them off when you need to.

The model has a tendency to try to sit flat on one of the dodecahedral side until it's finished so you will need to use a small folded piece of card to support the base as you build. This is one of the trickiest assemblages in this book, so go slowly and carefully.

Assembling the Model

Build the model from the dodecahedral base. That means working on each dodecahedral face embossed with a five pointed star. Begin by adding one side of the pentagon to the two adjacent points of the star (Fig. 1a). Then add the remaining points and sides to complete the pentagon (Fig. 1b).

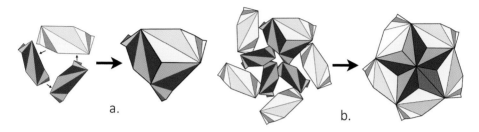

a.

b.

Fig. 1 creating the pentagon base

Now build up another five pentagons around the base (Fig. 2).

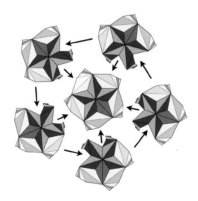

Fig. 2 adding the five pentagons to the base

Once you start building the second storey of pentagons, take care not to damage the joints in the base pentagon (try placing the model in a large bowl). Use sticky memos to hold the modules in place until you add the final piece and then carefully peel them off while holding the modules in place. When you have finished the model, you should be able to see the Small Triambic Icosahedron formed by the Unit B modules.

This model brings us to the end of the book, but I am already planning a further one exploring some of the wonderful stellations of the Rhombic Triacontahedron. Until then, I can only wish you happy folding.

More from Tarquin

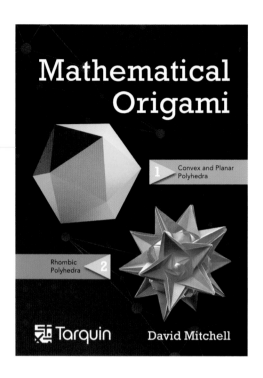

Mathematical Origami
Second Edition
David Mitchell

Each of the beautiful and fundamental mathematical shapes described in this book is achieved by folding sheets of standard A4 paper. It is remarkable what can been done and David Mitchell gives clear step by step instructions for each. He has gathered together a most impressive collection which will amaze and interest mathematics teachers and other admirers of pure geometrical forms. This second edition extends the models considerably - adding 60 new pages of activities. The models have been divided into Convex and Planar Polyhedra and Rhombic Polyhedra - using modular origami to the maximum effect.

Paperback   ISBN 9781911093039
Hardback   ISBN 9781911093169
Ebook   ISBN 9781911093480

Action Modular Origami
Tung Ken Lam

How to fold paper into geometric sculptures that move, spin and change

This is first and only book of action modular origami. All models are absorbing to make and rewarding to assemble, attractive to look at and also pleasing to play with. They spin, move and change shape in unique ways. Tung Ken Lam's clear instructions and approachable writing style will make this an instant classic, used by recreational origami enthusiasts, teachers and students, serious mathematicians and paper modellers.

Paperback   **ISBN 9781911093947**
Ebook   **ISBN 9781911093954**

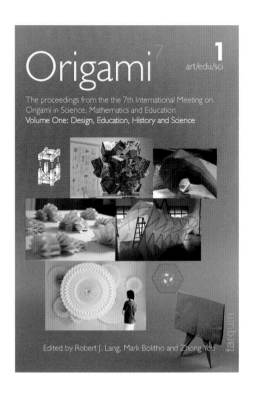

OSME[7] Proceedings
Volume 1: Education, Design, History and Science

Origami[7] is a collection of papers published for the 7th International Meeting on Origami in Science, Mathematics and Education (7OSME), held at Oxford University in the United Kingdom from September 4–7, 2018. 7OSME is the seventh conference in a series dedicated to research in the applications of origami and folding in the conference title fields, as well as in technology, design and history.

Volume 1:

Paperback   **ISBN 9781911093893**
Ebook   **ISBN 9781858118369**

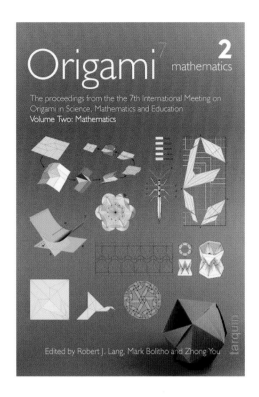

OSME[7] Proceedings
Volume 2: Mathematics

Origami[7] is a collection of papers published for the 7th International Meeting on Origami in Science, Mathematics and Education (7OSME), held at Oxford University in the United Kingdom from September 4–7, 2018. 7OSME is the seventh conference in a series dedicated to research in the applications of origami and folding in the conference title fields, as well as in technology, design and history.

Volume 2:

Paperback   **ISBN 9781911093909**
Ebook   **ISBN 9781858118376**

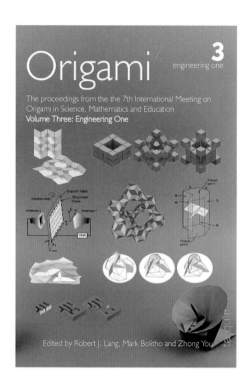

OSME[7] Proceedings
Volume 3: Engineering One

Origami[7] is a collection of papers published for the 7th International Meeting on Origami in Science, Mathematics and Education (7OSME), held at Oxford University in the United Kingdom from September 4–7, 2018. 7OSME is the seventh conference in a series dedicated to research in the applications of origami and folding in the conference title fields, as well as in technology, design and history.

Volume 3 (first of 2 volumes of engineering):

Paperback   **ISBN 9781911093916**
Ebook   **ISBN 9781858118383**

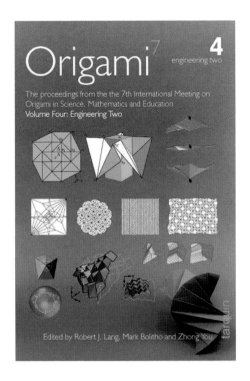

OSME[7] Proceedings
Volume 4: Engineering Two

Origami[7] is a collection of papers published for the 7th International Meeting on Origami in Science, Mathematics and Education (7OSME), held at Oxford University in the United Kingdom from September 4–7, 2018. 7OSME is the seventh conference in a series dedicated to research in the applications of origami and folding in the conference title fields, as well as in technology, design and history.

Volume 3 (first of 2 volumes of engineering):

Paperback   ISBN 9781911093923
Ebook   ISBN 9781858118390

Acknowledgements

Firstly I would like to thank Andrew Griffin at Tarquin Books for guiding me through this complex project, for his polite and friendly advice and support, and for believing in my work from the beginning. I must thank my sister, Madeleine Shepherd, who first suggested that I submit my work to Tarquin.

I must also give thanks to the team of beta-testing folders on Facebook, Gavin Higginbottom, Peter 'Wonko' Whitehouse (Peter has a great origami blog at http://www.wonko.info/365origami/), William Holt and Gustaw 'Nerfox' Kubala (Compound of Three Cubes), all of whom gave very helpful feedback. I would like to thank origami masters David Brill and David Mitchell (origamiheaven.com), too, for their useful advice and support proffered in our various email exchanges and online discussions.

Finally but by no means least, as a gesture of gratitude I would like to dedicate this book to my wife Maria, who patiently watched our home gradually filling up with hoards of invading multicoloured polyhedra during lockdown.

Tarquin Group
www.tarquingroup.com

Tarquin produces a wide range of mathematical and origami titles like
Mathematical Origami David Mitchell
Action Modular Origami - Tung Ken Lam
OSME7 Proceedings
and many more - some details on preceding pages.
See the website for more details - books and ebooks available worldwide